275 YEARS AT THE AMERICAN PHILOSOPHICAL SOCIETY

Essays by Benjamin Franklin

Transactions of the
American Philosophical Society 1786

Portrait of Benjamin Franklin by Jean Baptise Greuze, 1777. Oil on canvas, H: 39 in., W: 33¹/4 in. American Philosophical Society. Gift at the bequest of Lamont duPont Copeland, 1983.

Essays by Benjamin Franklin

Transactions of the American Philosophical Society 1786

Collected and edited by the American Philosophical Society

American Philosophical Society Press
Philadelphia

Transactions of the
American Philosophical Society
Held at Philadelphia
For Promoting Useful Knowledge
Volume 107, Part 3

ISBN: 978-1-60618-073-0

U.S. ISSN: 0065-9746

Library of Congress Cataloging-in-Publication Data

Names: Franklin, Benjamin, 1706-1790, author. I American Philosophical
    Society.
Title: Essays by Benjamin Franklin : transactions of the American
    Philosophical Society, 1786 / collected and edited by the American
    Philosophical Society.
Description: Philadelphia : American Philosophical Society Press, 2018. I
    Series: Transactions of the American Philosophical Society I Includes
    bibliographical references and index.
Identifiers: LCCN 2018049614 I ISBN 9781606180730 (pbk. : alk. paper)
Subjects: LCSH: Franklin, Benjamin, 1706-1790. I Statesmen—United
    States—Biography. I
Inventors—United States—Biography. I Inventions—United States—
    History—18th century.
Classification: LCC E302.6.F82 F73 2018 I DDC 973.3092 [B] —dc23
LC record available athttps://lccn.loc.gov/2018049614

On the cover: *Silhouette of Benjamin Franklin* by Beatrix Sherman. Paper.
American Philosophical Society. Gift of Sam Carrier and Susan Kane.

CONTENTS

N° V.

Letter to Mr. NAIRNE, of London, from Dr. Franklin, Proposing a Slowly Sensible Hygrometer for Certain Purposes

---

N° V.

Letter to Mr. NAIRNE, *of London.*

Paſſy, near Paris, Nov. 13th, 1780.

SIR,

Read Janua-
ry 28, 1786. THE qualities hitherto ſought in a hygrometer, or inſtrument to diſcover the degrees of
moiſture and dryneſs in the air, ſeem to have been, an
aptitude to receive humidity readily from a moiſt air, and
to part with it as readily to a dry air. Different ſubſtances
have been found to poſſeſs more or leſs of this quality;
but when we ſhall have found the ſubſtance that has it in
the greateſt perfection, there will ſtill remain ſome uncertainty in the concluſions to be drawn from the degree
ſhown by the inſtrument, ariſing from the actual ſtate of
the inſtrument itſelf as to heat and cold. Thus, if two
bottles or veſſels of glaſs or metal being filled, the one with
cold and the other with hot water, are brought into a room,
the moiſture of the air in the room will attach itſelf in
quantities to the ſurface of the cold veſſel, while if you
actually wet the ſurface of the hot veſſel, the moiſture will
immediately quit it, and be abſorbed by the ſame air.
And thus in a ſudden change of the air from cold to warm,
the inſtrument remaining longer cold may condenſe and
abſorb more moiſture, and mark the air as having become

G 2   more

SIR,

Read January 28, 1786 T HE qualities hitherto sought in a hygrometer, or instrument to discover the degrees of moisture and dryness in the air, seem to have been, an aptitude to receive humidity readily from a moist air, and to part with it as readily to a dry air. Different substances have been found to possess more or less of this quality; but when we shall have found the substance that has it in the greatest perfection, there will still remain some uncertainty in the conclusions to be drawn from the degree shown by the instrument, arising from the actual fate of the instrument itself as to heat and cold. Thus, if two bottles or vessels of glass or metal being filled, the one with cold and the other with hot water, are brought into a room, the moisture of the air in the room will attach itself in quantities to the surface of the cold vessel, while if you actually wet the surface of the hot vessel, the moisture will immediately quit it, and be absorbed by the same air. And thus in a sudden change of the air from cold to warm, the instrument remaining longer cold may condense and absorb more moisture, and mark the air as having become more humid than it is in reality, and the contrary in a change from warm to cold.

But if such a suddenly changing instrument could be freed from these imperfections, yet when the design is to discover the different degrees of humidity in the air of different countries, I apprehend the quick sensibility of the instrument to be rather a disadvantage; since, to draw the desired conclusions from it, a constant and frequent observation day and night in each country will be necessary for a year or years, and the mean of each different set of observations is to be found and determined. After all which some uncertainty will remain respecting the different degrees of exactitude with which different persons may have made and taken notes of their observations.

For these reasons, I apprehend that a substance which, though capable of being distended by moisture and contracted by dryness, is so slow in receiving and parting with its humidity that the frequent changes in the atmosphere have not time to affect it sensibly, and which therefore should gradually take nearly the medium of all those changes and preserve it constantly, would be the most proper substance of which to make such an hygrometer.

Such an instrument, you, my dear sir, though without intending it, have made for me; and I, without desiring or expelling it, have received from you. It is therefore with propriety that I address to you the following account of it; and the more, as you have both a head to contrive and a hand to execute the means of perfecting it. And I do this with greater pleasure, as it affords me the opportunity of renewing that ancient correspondence and acquaintance with you, which to me was always so pleasing and so instructive.

You may possibly remember, that in or about the year 1758, you made for me a set of artificial magnets, six in number, each five and a half inches long, half an inch broad, and one eighth of an inch thick. These, with two pieces of soft iron, which together equalled one of the magnets, were inclosed in a little box of mahogany wood, the grain of which ran with, and not across, the length of the box; and the box was closed by a little shutter of the same wood, the grain of which ran across the box; and the ends of this shutting piece were bevelled so as to fit and slide in a kind of dovetail groove when the box was to be shut or opened.

I had been of opinion that good mahogany wood was not affected by moisture so as to change its dimensions, and that it was always to be found as the tools of the workman left it. Indeed the difference at different times in the same country, is so small as to be scarcely in a common way observable. Hence the box which was made so as to allow sufficient room for the magnets to slide out and in freely, and, when in, afforded them so much play that by shaking the box one could make them strike the opposite sides alternately, continued in the same state all the time I remained in England, which was four years, without any apparent alteration. I left England in August 1762, and arrived at Philadelphia in October the same year. In a few weeks after my arrival, being desirous of showing your magnets to a philosophical friend, I found them so tight in the box, that it was with difficulty I got them out; and constantly during the two years I remained there, viz. till November 1764, this difficulty of getting them out and in continued. The little shutter too, as wood does not shrink length ways of the grain, was found too long to enter its grooves, and not being used, was mislaid and lost; and I afterwards had another made that fitted.

In December 1764 I returned to England, and after some time I observed that my box was become full big enough for my magnets, and too wide for my new shutter; which was so much too short for its grooves; that it was apt to fall out; and to make it keep in, I lengthened it by adding to each end a little coat of sealing-wax.

I continued in England more than ten years, and during all that time after the first change, I perceived no alteration. The magnets had the same freedom in their box, and the little shutter continued with the added sealing-wax to fit its grooves, till some weeks after my second return to America.

As I could not imagine any other cause for this change of dimensions in the box, when in the different countries, I concluded, first generally that the air of England was moister than that of America. And this I supposed an effect of its being an island, where every wind that blew must necessarily pass over some sea before it arrived, and of course lick up some vapour. I afterwards indeed doubted whether it might be just only so far as related to the city of London, where I resided; because there are many causes of moisture in the city air, which do not exist to the same degree in the country; such as the brewers and dyers boiling caldrons, and the great number of pots and teakettles continually on the fire, sending fourth abundance of vapour; and also the number of animals who by their breath continually increase it; to which may be added, that even the vast quantity of sea coals burnt there, do in kindling discharge a great deal of moisture.

When I was in England, the last time, you also made for me a little achromatic pocket telescope, the body was brass, and it had a round case, (I think of thin wood) covered with shagrin. All the while I remained in England, though possibly there might be some small changes in the dimensions of this case, I neither perceived nor suspected any. There was always comfortable room for the telescope to slip in and out. But soon after I arrived in America, which was in May 1775, the case became too small for the instrument, it was with much difficulty and various contrivances that I got it out, and I could never after get it in again, during my stay there, which was eighteen months. I brought it with me to Europe, but left the case as useless, imagining that I should find the continental air of France as dry as that of Pennsylvania, where my magnet box had also returned a second time to its narrowness, and pinched the pieces, as heretofore, obliging me too, to scrape the sealing-wax off the ends of the shutter.

I had not been long in France, before I was surprised to find, that my box was become as large as it had always been in England, the magnets entered and came out with the same freedom, and, when in, I could rattle them against its sides; this has continued to be the case without sensible variation. My habitation is out of Paris distant almost a league, so that the moist air of the city cannot be supposed to have much effect upon the box. I am on a high dry hill in a free air as likely to be dry as any air in France. Whence it seems probable that the air of England in general may as well as that of London, be moister than the air of America, since that of France is so, and in a part so distant from the sea.

The greater dryness of the air in America appears from some other observations. The cabinet work formerly sent us from London, which consisted in thin plates of fine wood glued upon fir, never would stand with us, the vaneering, as those plates are called, would get loose and come off; both woods shrinking, and their grains often crossing, they were forever cracking and flying. And in my electrical experiments there, it was remarkable, that a mahogany table on which my jars stood under the prime conductor to be charged, would often be so dry, particularly when the wind had been some time at north-west which with us is a very drying wind, as to isolate the jars, and prevent their being charged till I had formed a communication between their coatings and the earth. I had a like table in London which I used for the same purpose all the time I resided there; but it was never so dry as to refuse conducting the electricity.

Now what I would beg leave to recommend to you, is, that you would recollect, if you can, the species of mahogany of which you made my box, for you know there is a good deal of difference in woods that go under that name; or if that cannot be, that you would take a number of pieces of the closest and finest grained mahogany that you can meet with, plane them to the thinness of about a line, and the width of about two inches across the grain, and fix each of the pieces in some instrument that you can contrive, which will permit them to contract and dilate, and will show, in sensible degrees, by a moveable hand upon a marked scale, the otherwise less sensible quantities of such contraction and dilatation. If these instruments are all kept in the same place while making, and are graduated together while subject to the same degrees of moisture or dryness, I apprehend you will have so many comparable hygrometers, which being sent into different countries, and continued there for some time, will find and show there the mean of the different dryness and moisture of the air of those countries, and that with much less trouble than by any hygrometer hitherto in use.

With great esteem,

I am, dear sir,

Your most obedient,

And most humble servant,

B. FRANKLIN.

A Letter from Dr. B. FRANKLIN *to Dr.* INGENHAUSZ,
*Physician to the Emperor, at Vienna, on the Causes and Cures
of Smokey Chimneys*

TRANSACTIONS

OF THE

American PHILOSOPHICAL SOCIETY, *&c.*

N° I.

A Letter from Dr. B. FRANKLIN *to Dr.* INGENHAUSZ,
Phyſician to the Emperor, at Vienna.

Dear Friend,   At ſea, Auguſt 28th, 1785.

Read 21ſt
Oct. 1785.

IN one of your letters, a little before I left France, you deſire me to give you in writing my thoughts upon the conſtruction and uſe of chimneys, a ſubject you had ſometimes heard me touch upon in converſation. I embrace willingly this leiſure afforded by my preſent ſituation to comply with your re-queſt, as it will not only ſhow my regard to the deſires of a friend, but may at the ſame time be of ſome utility to others; the doctrine of chimneys appearing not to be as yet generally well underſtood, and miſtakes reſpecting them being attended with conſtant inconvenience, if not remedied; and with fruitleſs expence, if the true remedies are miſtaken.

Thoſe who would be acquainted with this ſubject ſhould begin by conſidering on what principle ſmoke aſcends in any chimney. At firſt many are apt to think that ſmoke

is

Dear Friend,   At sea, August 28th, 1785.

Read 21st   IN one of your letters, a little before I left France, you
Oct. 1785.   desire me to give you in writing my thoughts upon the
construction and use of chimneys, a subject you had sometimes
heard me touch upon in conversation. I embrace willingly this
leisure afforded by my present situation to comply with your
request, as it will not only show my regard to the desires of a
friend, but may at the same time be of some utility to others; the
doctrine of chimneys appearing not to be as yet generally well
understood, and mistakes respecting them being attended with
constant inconvenience, if not remedied; and with fruitless experience, if the true remedies are mistaken.

Those who would be acquainted with this subject should
begin by considering on what principle smoke ascends in any
chimney. At first many are apt to think that smoke is in its nature
and of itself specifically lighter than air, and rises in it for the
same reason that cork rises in water. These see no cause why
smoke should not rise in the chimney, though the room be ever
so close. Others think there is a power in chimneys to *draw* up
the smoke, and that there are different forms of chimneys that
afford more or less of this power. These amuse themselves with
searching for the best form. The equal dimensions of a funnel in
its whole length is not thought artificial enough, and it is made,
for fancied reasons, sometimes tapering and narrowing from below
upwards, and sometimes the contrary, &c. &c. A simple experiment
or two may serve to give more correct ideas. Having lit a pipe of
tobacco, plunge the stem to the bottom of a decanter half filled
with cold water; then putting a rag over the bowl, blow through it
and make the smoke descend in the stem of the pipe, from the
end of which it will rise in bubbles through the water; and being
thus cooled, will not afterwards rise to go out through the neck of
the decanter, but remain spreading itself and resting on the surface
of the water. This shows that smoke is really heavier than air, and
that it is carried upwards only when attached to, or acted upon,
by air that is heated, and thereby rarefied and rendered specifically
lighter than the air in its neighbourhood.

Editor's Note: All numbered figures appear together at the end of the article. See Plate I.

Smoke being rarely seen but in company with heated air, and its upward motion being visible, though that of the rarefied air that drives it is not so, has naturally given rise to the error.

I need not explain to you, my learned friend, what is meant by rarefied air; but if you make the public use you propose of this letter, it may fall into the hands of some who are unacquainted with the term and with the thing. These then may be told, that air is a fluid which has weight as well as others, though about eight hundred times lighter than water. That heat makes the particles of air recede from each other and take up more space, so that the same weight of air heated will have more bulk, than equal weights of cold air which may surround it, and in that case must rise, being forced upwards by such colder and heavier air, which presses to get under it and take its place. That air is so rarefied or expanded by heat, may be proved to their comprehension by a lank blown bladder, which laid before a fire will soon swell, grow tight and burst.

Another experiment may be to take a glass tube about an inch in diameter, and twelve inches long, open at both ends and fixed upright on legs so that it need not be handled, for the hands might warm it. At the end of a quill fasten five or six inches of the finest light filament of silk, so that it may be held either above the upper end of the tube or under the lower end, your warm hand being at a distance by the length of the quill. If there were any Plate 1 motion of air through the tube, it would manifest itself by Figure 1. its effect on the silk; but if the tube and the air in it are of the same temperature with the surrounding air, there will be no such motion, whatever may be the form of the tube, whether crooked or strait, narrow below and widening upwards, or the contrary; the air in it will be quiescent. Warm the tube, and you will find as long as it continues warm, a constant current of air entering below and passing up through it, till discharged at the top; because the warmth of the tube being communicated to the air it contains, rarefies that air and makes it lighter than the air without, which therefore presses in below, forces it upwards, follows and takes its place, and is rarefied in its turn. And, without warming the tube, if you hold under it a knob of hot iron, the air thereby heated will rise and fill the tube, going out at its top, and this motion in the tube will continue as long as the knob remains hot, because the air entering the tube below is heated and rarefied by passing near and over that knob.

That this motion is produced merely by the difference of specific gravity between the fluid within and that without the tube, and not by any fancied form of the tube itself, may appear by plunging it into water contained in a glass jar a foot deep, through which such motion might be seen. The water within and without the tube being of the same specific gravity, balance each other, and both remain at rest. But take out the tube, stop its bottom with a finger and fill it with olive oil, which is lighter than water, then stopping the top, place it as before, its lower end under water, its top a very little above. As long as you keep the bottom stops, the fluids remain at rest, but the moment it is unstopt, the heavier enters below, forces up the lighter, and takes its place. And the motion then ceases, merely because the new fluid cannot be successively made lighter, as air may be by a warm tube.

In fact, no form of the funnel of a chimney has any share in its operation or effect respecting smoke, except its height. The longer the funnel, if erect, the greater its force when filled with heated and rarefied air, to *draw* in below and drive up the smoke, if one may, in compliance with custom, use the expression *draw*, when in fact it is the superior weight of the surrounding atmosphere that *presses* to enter the funnel below, and so *drives up* before it the smoke and warm air it meets with in its passage.

I have been the more particular in explaining these first principles, because, for want of clear ideas respecting them, much fruitless expense has been occasioned; not only single chimneys, but in some instances, within my knowledge, whole stacks having been pulled down and rebuilt with funnels of different forms, imagined more powerful in *drawing* smoke; but having still the same height and the same opening below, have performed no better than their predecessors.

What is it then which makes a *smoky chimney*, that is, a chimney which instead of conveying up all the smoke, discharges a part of it into the room, offending the eyes and damaging the furniture?

The causes of this effect, which have fallen under my observation, amount to *nine*, differing from each other, and therefore requiring different remedies.

1. *Smoky chimneys in a new house, are such, frequently from mere want of air.* The workmanship of the rooms being all good, and just out of the workman's hand, the joints of the boards of the flooring, and of the panels of wainscotting are all true and tight, the more so as the walls, perhaps not yet thoroughly dry, preserve a dampness in the air of the room which keeps the wood-work swelled and close. The doors and the sashes too, being worked with truth, shut with exactness, so that the room is as tight as a snuff-box, no passage being left open for air to enter, except the key-hole, and even that is sometimes covered by a little dropping shutter. Now if smoke cannot rise but as connected with rarefied air, and a column of such air, suppose it filling the funnel, cannot rise, unless other air be admitted to supply its place; and if, therefore, no current of air enter the opening of the chimney, there is nothing to prevent the smoke coming out into the room. If the motion upwards of the air in a chimney that is freely supplied, be observed by the rising of the smoke or a feather in it, and it be considered that in the time such feather takes in rising from the fire to the top of the chimney, a column of air equal to the content of the funnel must be discharged, and an equal quantity supplied from the room below, it will appear absolutely impossible that this operation should go on if the tight room is kept shut; for were there any force capable of drawing constantly so much air out of it, it must soon be exhausted like the receiver of an air pump, and no animal could live in it. Those therefore who stop every crevice in a room to prevent the admission of fresh air, and yet would have their chimney carry up the smoke, require inconsistencies, and expect impossibilities. Yet under this situation, I have seen the owner of a new house, in despair, and ready to sell it for much less than it cost, conceiving it uninhabitable, because not a chimney in any one of its rooms would carry off the smoke, unless a door or window were left open. Much expense has also been made, to alter and amend new chimneys which had really no fault; in one house particularly that I knew, of a nobleman in Westminster, that expense amounted to no less than three hundred pounds, *after* his house had been, as he thought, finished and all charges paid. And after all, several of the alterations were ineffectual, for want of understanding the true principles.

Remedies. When you find on trial, that opening the door or a window, enables the chimney to carry up all the smoke, you may be sure that want of air *from without*, was the cause of its smoking. I say *from without*, to guard you against a common mistake of those who may tell you, the room is large, contains abundance of air, sufficient to supply any chimney, and therefore it cannot be that the chimney wants air. These reasoners are ignorant, that the largeness of a room, if tight, is in this case of small importance, since it cannot part with a chimney full of its air without occasioning so much vacuum; which it requires a great force to effect, and could not be borne if effected.

It appearing plainly, then, that some of the outward air must be admitted, the question will be, how much is *absolutely necessary*; for you would avoid admitting more, as being contrary to one of your intentions in having a fire, viz. that of warming your room. To discover this quantity, shut the door gradually while a middling fire is burning, till you find that, before it is quite shut, the smoke begins to come out into the room, then open it a little till you perceive the smoke comes out no longer. There hold the door, and observe the width of the open crevice between the edge of the door and the rabbit it should shut into. Suppose the distance to be half an inch, and the door eight feet high, you find thence that your room requires an entrance for air equal in area to ninety six half inches, or forty eight square inches, or a passage of six inches by eight. This however is a large supposition, there being few chimneys, that, having a moderate opening and a tolerable height of funnel, will not be satisfied with such a crevice of a quarter of an inch; and I have found a square of six by six, or thirty six square inches, to be a pretty good medium, that will serve for most chimneys. High funnels with small and low openings, may indeed be supplied through a less space, because, for reasons that will appear hereafter, the *forcé of levity*, if one may so speak, being greater in such funnels, the cool air enters the room with greater velocity, and consequently more enters in the same time. This however has its limits, for experience shows that no increased velocity so occasioned, has made the admission of air through the key-hole equal in quantity to that through an open door; though through the door the current moves slowly, and through the key-hole with great rapidity.

It remains then to be considered how and where this necessary quantity of air from without is to be admitted so as to be least inconvenient. For, if at the door, left so much open, the air thence proceeds directly to the chimney, and in its way comes cold to your back and heels as you sit before your fire. If you keep the door shut, and raise a little the sash of your window, you feel the same inconvenience. Various have been the contrivances to avoid this, such as bringing in fresh air through pipes in the jams of the chimney, which pointing upwards should blow the smoke up the funnel; opening passages into the funnel above, to let in air for the same purpose. But these produce an effect contrary to that intended: For as it is the constant current of air passing from the room *through the opening of the chimney* into the funnel which prevents the smoke coming out into the room, if you supply the funnel by other means or in other ways with the air it wants, and especially if that air be cold, you diminish the force of that current, and the smoke in its efforts to enter the room finds less resistance.

The wanted air must then *indispensably* be admitted into the room, to supply what goes off through the opening of the chimney. M. Gauger, a very ingenious and intelligent French writer on the subject, proposes with judgment to admit it *above* the opening of the chimney; and to prevent inconvenience from its coldness, he directs its being made to pass in its entrance through winding cavities made behind the iron back and sides of the fireplace, and under the iron hearth-plate; in which cavities it will be warmed, and even heated, so as to contribute much, instead of cooling, to the warming of the room. This invention is excellent in itself, and may be used with advantage in building new houses; because the chimneys may then be so disposed, as to admit conveniently the cold air to enter such passages: But in houses built without such views, the chimneys are often so situated, as not to afford that convenience, without great and expensive alterations. Easy and cheap methods, though not quite so perfect in themselves, are of more general utility; and such are the following.

In all rooms where there is a fire, the body of air warmed and rarefied before the chimney is continually changing place, and making room for other air that is to be warmed in its turn. Part of it enters and goes up the chimney, and the rest rises and takes place near the ceiling. If the room be lofty, that warm air remains above our heads as long as it continues warm, and we are little benefited by it, because it does not descend till it is cooler. Few can imagine the difference of climate between the upper and lower parts of such a room, who have not tried it by the thermometer, or by going up a ladder till their heads are near the ceiling. It is then among this warm air that the wanted quantity of outward air is best admitted, with which being mixed, its coldness is abated, and its inconvenience diminished so as to become scarce observable. This may be easily done, by drawing down about an inch the upper sash of a window; or, if not moveable, by cutting such a crevice through its frame; in both which cases, it will be well to place a thin shelf of the length, to conceal the opening, and sloping upwards to direct the entering air horizontally along and under the ceiling. In some houses the air may be admitted by such a crevice made in the wainscot, cornish or plastering, near the ceiling and over the opening of the chimney. This, if practicable, is to be chosen, because the entering cold air will there meet with the warmest rising air from before the fire, and be soonest tempered by the mixture. The same kind of shelf should also be placed here. Another way, and not a very difficult one, is to take out an upper pane of glass in one of your sashes, set it in a tin frame, giving it two springing angular sides, and then Plate 1. replacing it, with hinges below on which it may be turned Figure 2. to open more or less above. It will then have the appearance of an internal skylight. By drawing this pane in, more or less, you may admit what air you find necessary. Its position will naturally throw that air up and along the ceiling. This is what is called in France a *Was ist das?* As this is a German question, the invention is probably of that nation, and takes its name from the frequent asking of that question when it first appeared. In England, some have of late years cut a round hole about five inches diameter in a pane of the sash and placed against it a circular plate of tin hung on an axis, and cut into vanes, which being separately bent a little obliquely, are acted upon by the entering air, so as to force the plate continually round like the vanes of a windmill. This admits the outward air, and by the continual whirling of the vanes, does in some degree disperse it. The noise only, is a little inconvenient.

2. A second cause of the smoking of chimneys is, *their openings in the room being too large*; that is, too wide, too high or both.

Architects in general have no other ideas of proportion in the opening of a chimney, than what relate to symmetry and beauty, respecting the dimensions of the room*; while its true proportion, respecting its function and utility depends on quite other principles; and they might as properly proportion the step in a staircase to the height of the story, instead of the natural elevation of men's legs in mounting. The proportion then to be regarded, is what relates to the height of the funnel. For as the funnels in the different stories of a house are necessarily of different heights or lengths, that from the lowest floor being the highest or longest, and those of the other floors shorter and shorter, till we come to those in the garrets, which are of course the shortest; and the force of draft being, as already said, in proportion to the height of funnel filled with rarefied air; and a current of air from the room into the chimney, sufficient to fill the opening, being necessary to oppose and prevent the smoke coming out into the room; it follows that the openings of the longest funnels may be larger, and that those of the shorter funnels should be smaller. For if there be a large opening to a chimney that does not draw strongly, the funnel may happen to be furnished with the air it demands by a partial current entering on one side of the opening, and leaving the other side free of any opposing current, may permit the smoke to issue there into the room. Much too of the force of draft in a funnel depends on the degree of rarefaction in the air it contains, and that depends on the nearness to the fire of its passage in entering the funnel. If it can enter far from the fire on each side, or far above the fire, in a wide or high opening, it receives little heat in passing by the fire, and the contents of the funnel is by that means less different in levity from the surrounding atmosphere, and its force in drawing consequently weaker. Hence if too large an opening be given to chimneys in upper rooms, those rooms will be smoky: On the other hand, if too small openings be given to chimneys in the lower rooms, the entering air operating too directly and violently on the fire, and afterwards strengthening the draft as it ascends the funnel, will consume the fuel too rapidly.

Remedy. As different circumstances frequently mix themselves in these matters, it is difficult to give precise dimensions

---

*See Appendix, N° I.

for the openings of all chimneys. Our fathers made them generally much too large; we have lessened them; but they are often still of greater dimension than they should be, the human eye not being easily reconciled to sudden and great changes. If you suspect that your chimney smokes from the too great dimension of its opening, contract it by placing moveable boards so as to lower and narrow it gradually, till you find the smoke no longer issues into the room. The proportion so found will be that which is proper for that chimney, and you may employ the bricklayer or mason to reduce it accordingly. However, as, in building new houses, something must be sometimes hazarded, I would make the openings in my lower rooms about thirty inches square and eighteen deep, and those in the upper, only eighteen inches square and not quite so deep; the intermediate ones diminishing in proportion as the height of funnel diminished. In the larger openings, billets of two feet long, or half the common length of cordwood, may be burnt conveniently; and for the smaller, such wood may be sawed into thirds. Where coals are the fuel, the grates will be proportioned to the openings. The same depth is nearly necessary to all, the funnels being all made of a size proper to admit a chimney-sweeper. If in large and elegant rooms custom or fancy should require the appearance of a larger chimney, it may be formed of expensive marginal decorations, in marble, &c. In time perhaps that which is fitted in the nature of things, may come to be thought handsomest. But at present when men and women in different countries show themselves dissatisfied with the forms God has given to their heads, waists and feet, and pretend to shape them more perfectly, it is hardly to be expected that they will be content always with the best form of a chimney. And there are some I know so bigotted to the fancy of a large noble opening, that rather than change it, they would submit to have damaged furniture, sore eyes and skins almost smoked to bacon.

3. Another cause of smoky chimneys is, *too short a funnel.* This happens necessarily in some cases, as where a chimney is required in a low building; for, if the funnel be raised high above the roof, in order to strengthen its draft, it is then in danger of being blown down, and crushing the roof in its fall.

Remedies. Contract the opening of the chimney, so as to oblige all the entering air to pass through or very near the fire; whereby it will be more heated and rarefied, the funnel itself be more warmed, and its contents have more of what may be called the force of levity, so as to rise strongly and maintain a good draft at the opening.

Or you may in some cases, to advantage, build additional stories over the low building, which will support a high funnel.

If the low building be used as a kitchen, and a contraction of the opening therefore inconvenient, a large one being necessary, at least when there are great dinners, for the free management of so many cooking utensils in such case I would advise the building of two more funnels joining to the first, and having three moderate openings, one to each funnel, instead of one large one. When there is occasion to use but one, the other two may be kept shut by sliding plates, hereafter to be described*; and two or all of them may be used together when wanted. This will indeed be an expense, but not an useless one, since your cooks will work with more comfort, see better than in a smoky kitchen what they are about, your visuals will be cleaner dressed and not taste of smoke, as is often the case; and to render the effect more certain, a stack of three funnels may be safely built higher above the roof than a single funnel.

The case of too short a funnel is more general than would be imagined, and often found where one would not expect it. For it is not uncommon, in ill-contrived buildings, instead of having a funnel for each room or fireplace, to bend and turn the funnel of an upper room so as to make it enter the side of another funnel that comes from below. By this means the upper room funnel is made short of course, since its length can only be reckoned from the place where it enters the lower room funnel; and that funnel is also shortened by all the distance between the entrance of the second funnel and the top of the stack: For all that part being readily supplied with air through the second funnel, adds no strength to the draft, especially as that air is cold when there is no fire in the second chimney. The only easy remedy here is, to keep the opening shut of that funnel in which there is no fire.

---

*See Appendix, N° II.

4. Another very common cause of the smoking of chimneys, is, *their overpowering one another.* For instance, if there be two chimneys in one large room, and you make fires in both of them, the doors and windows close shut, you will find that the greater and stronger fire shall overpower the weaker, and draw air down its funnel to supply its own demand; which air descending in the weaker funnel will drive down its smoke, and force it into the room. If, instead of being in one room, the two chimneys are in two different rooms, communicating by a door, the case is the same whenever that door is open. In a very tight house, I have known a kitchen chimney on the lowest floor, when it had a great fire in it, overpower any other chimney in the house, and draw air and smoke into its room, as often as the door was opened communicating with the staircase.

Remedy. Take care that every room have the means of supplying itself from without, with the air its chimney may require, so that no one of them may be obliged to borrow from another, nor under the necessity of lending. A variety of these means have been already described.

5. Another cause of smoking is, *when the tops of chimneys are commanded by higher buildings, or by a hill,* so that the wind blowing over such eminences falls like water over a dam, sometimes almost perpendicularly on the tops of the chimneys that lie in its way, and beats down the smoke contained in them.

Remedy. That commonly applied to this case, is a turncap made of tin or plate iron, covering the chimney above and on three sides, open on one side, turning on a spindle, and which being guided or governed by a vane, always presents its back to the current. This I believe may be generally effectual, though not certain, as there may be cases in which it will not succeed. Raising your funnels if practicable, so as their tops may be higher, or at least equal with the commanding eminence, is more to be depended on. But the turning cap, being easier and cheaper, should first be tried. If obliged to build in such a situation, I would choose to place my doors on the side next the hill, and the backs of my chimneys on the furthest side; for then the column of air falling over the eminence, and of course pressing on that below and forcing it to enter the doors, or *Was-ist-dase*s on that side, would tend to balance the pressure down the chimneys, and leave the funnels more free in the exercise of their functions.

6. There is another case of command, the reverse of that last mentioned. It is where the commanding eminence is farther from the wind than the chimney commanded. To explain this a figure may be necessary. Suppose then a building whose side A, happens to be exposed to the wind, and forms a kind of dam against its progress. The air obstructed by this dam will like water press and search for passages through it; and finding the top of   Plate 1. the chimney B, below the top of the dam, it will force it-   Figure 3. self down that funnel, in order to get through by some door or window open on the other side of the building. And if there be a fire in such chimney, its smoke is of course beat down, and fills the room.

Remedy. I know of but one, which is to raise such funnel higher than the roof, supporting it, if necessary, by iron bars. For a turncap in this case has no effect, the dammed up air pressing down through it in whatever position the wind may have placed its opening.

I know a city in which many houses are rendered smoky by this operation. For their kitchens being built behind, and connected by a passage with the houses, and the tops of the kitchen chimneys lower than the top of the houses, the whole side of a street when the wind blows against its back, forms such a dam as above described; and the wind so obstructed forces down those kitchen chimneys, (especially when they have but weak fires in them) to pass through the passge and house, into the street. Kitchen chimneys so formed and situated, have another inconvenience. In summer, if you open your upper room windows for air, a light breeze blowing over your kitchen chimney towards the house, though not strong enough to force down its smoke as aforesaid, is sufficient to waft it into your windows, and fill the rooms with it; which, besides the disagreeableness, damages your furniture.

7. Chimneys, otherwise drawing well, are sometimes made to smoke *by the improper and inconvenient situation of a door*. When the door and chimney are on the same side of the room as in the figure, if the door A, being in the corner is made to open against

the wall, which is common, as being there, when open, more out of
Plate 1. the way, it follows, that when the door is only opened in
Figure 4. part, a current of air ruffling in passes along the wall into
and across the opening of the chimney B, and flirts some of the
smoke out into the room. This happens more certainly when the
door is shutting, for then the force of the current is augmented,
and becomes very inconvenient to those who, warming themselves
by the fire, happen to fit in its way.

The *Remedies* are obvious and easy. Either put an intervening
screen from the wall round great part of the fireplace; or, which
is perhaps preferable, shift the hinges of your door, so as it may
open the other way, and when open throw the air along the
other wall.

8. A room that has no fire in its chimney, is sometimes filled
with *smoke which is received at the top of its funnel and descends
into the room.* In a former paper* I have already explained the
descending currents of air in cold funnels; it may not be amiss
however to repeat here, that funnels without fires have an effect
according to their degree of coldness or warmth, on the air that
happens to be contained in then. The surrounding atmosphere is
frequently changing its temperature; but stacks of funnels covered
from winds and sun by the house that contains them, retain a more
equal temperature. If, after a warm season, the outward air suddenly
grows cold, the empty warm funnels begin to draw strongly upward;
that is, they rarefy the air contained in them, which of course
rises, cooler air enters below to supply its place, is rarefied in its
turn and rises; and this operation continues, till the funnel grows
cooler, or the outward air warmer, or both, when the motion ceases.
On the other hand, if after a cold season, the outward air suddenly
grows warm and of course lighter, the air contained in the cool
funnels, being heavier, descends into the room; and the warmer
air which enters their tops being cooled in its turn, and made
heavier, continues to descend; and this operation goes on, till the
funnels are warmed by the passing of warm air through them, or
the air itself grows cooler. When the temperature of the air and
of the funnels is nearly equal, the difference of warmth in the air

*See Appendix, N° II.

between day and night is sufficient to produce these currents, the air will begin to ascend the funnels as the cool of the evening comes on, and this current will continue till perhaps nine or ten o'clock the next morning, when it begins to hesitate; and as the heat of the day approaches, it sets downwards, and continues so till towards evening, when it again hesitates for some time, and then goes upwards constantly during the night, as before mentioned. Now when smoke issuing from the tops of neighboring funnels pales over the tops of funnels which are at the time drawing downwards, as they often are in the middle part of the day, such smoke is of necessity drawn into these funnels, and descends with the air into the chamber.

The *Remedy* is to have a sliding plate, hereafter described*, that will shut perfectly the offending funnel.

9. Chimneys which generally draw well, do nevertheless sometimes give smoke into the rooms, *it being driven down by strong winds passing over the tops of their funnels*, though not descending from any commanding eminence. This case is most frequent where the funnel is short, and the opening turned from the wind. It is the more grievous, when it happens to be a cold wind that produces the effect, because when you most want your fire, you are sometimes obliged to extinguish it. To understand this, it may be considered that the rising light air, to obtain a free issue from the funnel, must push out of its way or oblige the air that is over it to rise. In a time of calm or of little wind this is done visibly, for we see the smoke that is brought up by that air rise in a column above the chimney. But when a violent current of air, that is, a strong wind, passes over the top of a chimney, its particles have received so much force, which keeps them in a horizontal direction and follow each other so rapidly, that the rising light air has not strength sufficient to oblige them to quit that direction and move upwards to permit its issue. Add to this, that some of the current passing over that side of the funnel which it first meets with, viz, at A, having been compressed by the resis-
Plate 1. tance of the funnel, may expand itself over the flue, and
Figure 5. strike the interior opposite side at B, from whence it may be reflected downwards and from side to side in the direction of the pricked lines c c c.

---

*See Appendix, N° II.

Remedies. In some places, particularly in Venice, where they have not stacks of chimneys but single flues, the custom is, to Plate 1. open or widen the top of the flue rounding in the true form Figure 6. of a funnel; which some think may prevent the effect just mentioned, for that the wind blowing over one of the edges into the funnel may be slanted out again on the other side by its form. I have had no experience of this; but I have lived in a windy country, where the contrary is practiced, the tops of the flues being narrowed inwards, so as to form a slit for the issue of the smoke, long as the breadth of the funnel, and only four inches wide. This seems to have been contrived on a supposition that the entry of the wind would thereby be obstructed, and perhaps it might have been imagined, that the whole force of the rising warm air being condensed, as it were, in the narrow opening, would thereby be strengthened, so as to overcome the resistance of the wind. This however did not always succeed; for when the wind was at north-east and blew fresh, the smoke was forced down by fits into the room I commonly sat in, so as to oblige me to shift the fire into another. The position of the slit of this funnel was indeed north-east and south-west. Perhaps if it had lain across the wind, the effect might have been different, But on this I can give no certainty. It seems a matter proper to be referred to experiment. Possibly a turn-cap might have been serviceable, but it was not tried.

Chimneys have not been long in use in England. I formerly saw a book printed in the time of queen Elizabeth, which remarked the then modern improvements of living, and mentioned among others the convenience of chimneys. "Our forefathers," said the author, "had no chimneys. There was in each dwelling house only one place for a fire, and the smoke went out through a hole in the roof; but now there is scarce a gentleman's house in England that has not at least one chimney in it." When there was but one chimney, its top might then be opened as a funnel, and perhaps, borrowing the form from the Venetians, it was then the flue of a chimney got that name. Such is now the growth of luxury, that in both England and France we must have a chimney for every room, and in some houses every possessor of a chamber, and almost every servant, will have a fire; so that the flues being necessarily built in stacks, the opening of each as a funnel is impracticable. This change of manners soon consumed the firewood of England, and will soon render fuel extremely scarce and dear in France, if

the use of coals be not introduced in the latter, kingdom as it has
been in the former, where it at first met with opposition; for there
is extant in the records of one of queen Elizabeth's Parliaments,
a motion made by a member, reciting, "that many dyers, brewers,
smiths, and other artificers of London, had of late taken to the use
of pitcoal for their fires, instead of wood, which filled the air
with noxious vapours and smoke, very prejudicial to the health,
particularly of persons coming out of the country; and therefore
moving that a law might pass to prohibit the use of such fuel (at
least during the session of parliament) by those artificers."—It
seems it was not then commonly used in private houses. Its supposed unwholesomeness was an objection. Luckily the inhabitants
of London have got over that objection, and now think it rather
contributes to render their air salubrious, as they have had no
general penitential disorder since the general use of coals, when,
before it, such were frequent. Paris still burns wood at an enormous
expense continually augmenting, the inhabitants having still that
prejudice to overcome. In Germany you are happy in the use of
stoves, which save fuel wonderfully: Your people are very ingenious
in the management of fire; but they may still learn something in
that art from the Chinese*, whose country being greatly populous
and fully cultivated, has little room left for the growth of wood,
and having not much other fuel that is good, have been forced
upon many inventions during a course of ages, for making a little
fire go as far as possible.

I have thus gone through all the common causes of the smoking of chimneys that I can at present recollect as having fallen
under my observation; communicating the remedies that I have
known successfully used for the different cases, together with the
principles on which both the disease and the remedy depend, and
confessing my ignorance wherever I have been sensible of it. You
will do well, if you publish, as you propose, this letter, to add in
notes, or as you please, such observations as may have occurred
to your attentive mind; and if other philosophers will do the same,
this part of science, though humble, yet of great utility, may in
time be perfected. For many years past, I have rarely met with a

---

*See Appendix, Nº III.

case of a smoky chimney, which has not been solvable on these principles, and cured by these remedies, where people have been willing to apply them; which is indeed not always the case; for many have prejudices in favor of the nostrums of pretending chimney-doctors and fumists, and some have conceits and fancies of their own, which they rather choose to try, than to lengthen a funnel, alter the size of an opening, or admit air into a room, however necessary; for some are as much afraid of fresh air as persons in the hydrophobia are of fresh water. I myself had formerly this prejudice, this *aerophobia*, as I now account it, and dreading the supposed dangerous effects of cool air, I considered it as an enemy, and closed with extreme care every crevice in the rooms I inhabited. Experience has convinced me of my error. I now look upon fresh air as a friend: I even sleep with an open window. I am persuaded that no common air from without, is so unwholesome as the air within a close room that has been often breathed and not changed. Moist air too, which formerly I thought pernicious, gives me now no apprehensions: For considering that no dampness of air applied to the outline of my skin, can be equal to what is applied to and touches it within, my whole body being full of moisture, and finding that I can lie two hours in a bath twice a week, covered with water, which certainly is much damper than any air can be, and this for years together, without catching cold, or being in any other manner disordered by it, I no longer dread mere moisture, either in air or in sheets or shirts: And I find it of importance to the happiness of life, the being freed from vain terrors, especially of objects that we are every day exposed inevitably to meet with. You physicians have of late happily discovered, after a contrary opinion had prevailed some ages, that fresh and cool air does good to persons in the small pox and other fevers. It is to be hoped that in another century or two we may all find out, that it is not bad even for people in health. And as in moist air, here I am at this present writing in a ship with above forty persons, who have had no other but moist air to breathe for six weeks past; every thing we touch is damp, and nothing dries, yet we are all as healthy as we should be on the mountains of Switzerland, whose inhabitants are not more so than those of Bermuda or St. Helena, islands on whose rocks the waves are dallied into millions of particles, which fill the air with damp, but produce no diseases, the moisture being pure, unmixed with the poisonous vapors arising from putrid marshes and stagnant pools, in which many insects die and corrupt the water. These places only, in my opinion, (which however I submit to yours) afford unwholesome air; and that it is not the mere water contained in damp air, but the volatile particles of corrupted animal matter mixed with that water, which renders such air pernicious to those who breathe it. And I imagine it a cause of the same kind that renders the air in close rooms, where the perspirable matter is breathed over and over again by a number of assembled people, so hurtful to health. After being in such a situation, many find themselves affected by that *febricula*, which the English alone call *a cold*, and, perhaps from the name, imagine that they caught the malady by *going out* of the room, when it was in fact by being in it.

You begin to think that I wander from my subject, and go out of my depth. So I return again to my chimneys.

We have of late many lecturers in experimental philosophy. I have wished that some of them would study this branch of that science, and give experiments in it as a part of their lectures. The addition to their present apparatus need not be very expensive. A number of little representations of rooms composed each of five panes of fash glass, framed in wood at the corners, with proportionable doors, and moveable glass chimneys, with openings of different sizes, and different lengths of funnel, and some of the rooms so contrived as to communicate on occasion with others, so as to form different combinations, and exemplify different cases; with quantities of green wax taper cut into pieces of strong fire for a little glass chimney, and blown out would continue to burn and give smoke as long as desired. With such an apparatus all the operations of smoke and rarefied air in rooms and chimneys might be seen through their transparent sides; and the effect of winds on chimneys, commanded or otherwise, might be shown by letting the entering air blow upon them through an opened window of the lecturer's chamber, where it would be constant while he kept a good fire in his chimney. By the help of such lectures our fumists would become better instructed. At present they have generally but one remedy, which perhaps they have known effectual in some one case of smoky chimneys, and they apply that indiscriminately to all the other cases, without success—but not without expense to their employers.

With all the science, however, that a man shall suppose himself possessed of in this article, he may sometimes meet with cases that shall puzzle him. I once lodged in a house at London, which, in a little room, had a single chimney and funnel. The opening was very small, yet it did not keep in the smoke, and all attempts to have a fire in this room were fruitless. I could not imagine the reason, till at length observing that the chamber over it, which had no fire-place in it, was always filled with smoke when a fire was kindled below, and that the smoke came through the cracks and crevices of the wainscot; I had the wainscot taken down, and discovered that the funnel which went up behind it, had a crack many feet in length, and wide enough to admit my arm, a breach very dangerous with regard to fire, and occasioned probably by an apparent irregular settling of one side of the house. The air entering this breach, freely, destroyed the drawing force of the funnel. The remedy would have been, filling up the breach or rather rebuilding the funnel: But the landlord rather chose to stop up the chimney.

Another puzzling case I met with at a friend's country house near London. His best room had a chimney in which, he told me, he never could have a fire, for all the smoke came out into the room. I flattered myself I could easily find the cause, and prescribe the cure. I had a fire made there, and found it as he said. I opened the door, and perceived it was not want of air. I made a temporary contraction of the opening of the chimney, and found that it was not its being too large that caused the smoke to issue. I went out and looked up at the top of the chimney: Its funnel was joined in the same stack with others, some of them shorter, that drew very well, and I saw nothing to prevent its doing the same. In fine, after every other examination I could think of, I was obliged to own the insufficiency of my skill. But my friend, who made no pretension to such kind of knowledge, afterwards discovered the cause himself. He got to the top of the funnel by a ladder, and looking down, found it filled with twigs and straw cemented by earth, and lined with feathers. It seems the house, after being built, had stood empty some years before he occupied it; and he concluded that some large birds had taken the advantage of its retired situation to make their nest there. The rubbish, considerable in quantity, being removed, and the funnel cleared, the chimney drew well, and gave satisfaction.

In general, smoke is a very tractable thing, easily governed and directed when one knows the principles and is well informed of the circumstances. You know I made it *descend* in my Pennsylvania stove. I formerly had a more simple construction, in which the same effect was produced, but visible to the eye. It was composed Plate 1. of two plates A B and C D, placed as in the figure. The Figure 7. lower plate A B rested with its edge in the angle made by the hearth with the back of the chimney. The upper plate was fixed to the breast, and lapt over the lower about six inches, leaving a space of four inches wide and the length of the plates (near two feet) between them. Every other passage of air into the funnel was well stopped. When therefore a fire was made at E, for the first time with charcoal, till the air in the funnel was a little heated through the plates, and then wood laid on, the smoke would rise to A, turn over the edge of that plate, descend to D, then turn under the edge of the upper plate, and go up the chimney. It was pretty to see, but of no great use. Placing therefore the under plate in a higher situation, I removed the upper plate C D, and placed Plate 1. it perpendicularly, so that the upper edge of the lower plate Figure 8. A B came within about three inches of it, and might be pushed farther from it, or suffered to come nearer to it by a moveable wedge between them. The flame then ascending from the fire at E, was carried to strike the upper plate, made it very hot, and its heat rose and spread with the rarefied air into the room.

I believe you have seen in use with me, the contrivance of a sliding-plate over the fire, seemingly placed to oppose the rising of the smoke, leaving but a small passage for it, between the edge of the plate and the back of the chimney. It is particularly described, and its uses explained, in my former printed letter, and I mention it here only as another instance of the tractability of smoke*.

---

*See Appendix, Nº II.

What is called the Staffordshire chimney, affords an example of the same kind. The opening of the chimney is bricked up, even with the fore-edge of its jams, leaving open only a passage over the grate of the same width, and perhaps eight inches high. The grate consists of semicircular bars, their upper bar of the greatest diameter, the others under it smaller and smaller, so that it has the appearance of half a round basket. It is, with the coals it contains, wholly without the wall that shuts up the chimney, yet the smoke bends and enters the passage above it, the draft being strong, because no air can enter that is not obliged to pass near or through the fire, so that all that the funnel is filled with is much heated, and of course much rarefied.

STAFFORDSHIRE FIRE-PLACE.

SIDE VIEW.   FRONT VIEW.

Much more of the prosperity of a winter country depends on the plenty and cheapness of fuel, than is generally imagined. In travelling I have observed, that in those parts where the inhabitants can have neither wood nor coal nor turf but at excessive prices, the working people live in miserable hovels, are ragged, and have nothing comfortable about them. But where fuel is cheap, (or where they have the art of managing it to advantage) they are well furnished with necessaries, and have decent habitations. The obvious reason is, that the working hours of such people are the profitable hours, and they who cannot afford sufficient fuel have fewer such hours in the twenty four, than those who have it cheap and plenty: For much of the domestic work of poor women, such as spinning, sewing, knitting; and of the men in those manufactures that require little bodily exercise, cannot well be performed where the fingers are numbed with cold: Those people, therefore, in cold weather, are induced to go to bed sooner, and lie longer in a morning, than they would do if they could have good fires or warm stoves to sit by; and their hours of work are not sufficient to produce the means of comfortable subsistence. Those public works, therefore, such as roads, canals, &c. by which fuel may be brought cheap into such countries from distant places, are of great utility; and those who promote them may be reckoned among the benefactors of mankind.

I have great pleasure in having thus complied with your request, and in the reflection that the friendship you honor me with, and in which I have ever been so happy, has continued so many years without the smallest interruption. Our distance from each other is now augmented, and nature must soon put an end to the possibility of my continuing our correspondence: But if consciousness and memory remain in a future state, my esteem and respect for you, my dear friend, will be everlasting.

B. F.

A P P E N D I X.

NOTES FOR THE LETTER UPON CHIMNEYS.

N° I.

THE latest work on architecture that I have seen, is that entitled NUTSHELLS, which appears to be written by a very ingenious man, and contains a table of the proportions of the openings of chimneys; but they relate solely to the proportions he gives his rooms, without the smallest regard to the funnels. And he remarks, respecting those proportions, that they are similar to the harmonic divisions of a monochord*. He does not indeed lay much stress on this; but it shows that we like the appearance of principles; and where we have not true ones, we have some satisfaction in producing such as are imaginary.

N° II.

THE description of the sliding plates here promised, and which hath been since brought into use under various names, with some immaterial changes, is contained in a former letter to J. B. Esq. as follows:

To J. B. Esq. at Boston, in New-England.

Dear Sir,   London, Dec. 2, 1758.

I HAVE executed here an easy simple contrivance, that I have long since had in speculation, for keeping rooms warmer in cold weather than they generally are, and with less fire. It is this. The opening of the chimney is contracted, by brick-work faced

---

*"It may be just remarked here, that upon comparing these proportions with those arising from the common divisions of the monochord, it happens that the first answers to unison, and although the second is a discord, the third answers to the third minor, the fourth to the third major, the fifth to the fourth, the filth to the fifth, and the seventh to the octave."

NUTSHELLS, page 85.

with marble slabs, to about two feet between the jams, and the breast brought down to within about three feet of the hearth.—An iron frame is placed just under the breast, and extending quite to the back of the chimney, so that a plate of the same metal may slide horizontally backwards and forwards in the grooves on each side of the frame. This plate is just so large as to fill the whole space, and shut the chimney entirely when thrust quite in, which is convenient when there is no fire. Drawing it out, so as to leave a space between its further edge and the back, of about two inches; this space is sufficient for the smoke to pass; and so large a part of the funnel being stopt by the rest of the plate, the passage of warm air out of the room, up the chimney, is obstructed and retarded, and by that means much cold air is prevented from coming in through crevices, to supply its place. This effect is made manifest three ways. First, when the fire burns briskly in cold weather, the howling or whistling noise made by the wind, as it enters the room through the crevices, when the chimney is open as usual, ceases as soon as the plate is slid in to its proper distance. Secondly, opening the door of the room about half an inch, and holding your hand against the opening, near the top of the door, you feel the cold air coming in against your hand, but weakly, if the plate be in. Let another person suddenly draw it out, so as to let the air of the room go up the chimney, with its usual freedom where chimneys are open, and you immediately feel the cold air rushing in strongly. Thirdly, if something be set against the door, just sufficient, when the plate is in, to keep the door nearly shut, by resisting the pressure of the air that would force it open: Then, when the plate is drawn out, the door will be forced open by the increased pressure of the outward cold air endeavoring to get in to supply the place of the warm air, that now passes out of the room to go up the chimney. In our common open chimneys, half the fuel is wasted, and its effect lost; the air it has warmed being immediately drawn off. Several of my acquaintance having seen this simple machine in my room, have imitated it at their own houses, and it seems likely to become pretty common. I describe it thus particularly to you, because I think it would be useful in *Boston*, where firing is often dear.

Mentioning chimneys puts me in mind of a property I formerly had occasion to observe in them, which I have not found taken notice of by others; it is, that in the summer time, when no fire is made in the chimneys, there is, nevertheless, a regular draft of air through them; continually passing upwards, from about five or six o'clock in the afternoon, till eight or nine o'clock the next morning, when the current begins to slacken and hesitate a little, for about half an hour, and then sets as strongly down again, which it continues to do till towards five in the afternoon, then slackens and hesitates as before, going sometimes a little up, then a little down, till in about a half an hour it gets into a steady upward current for the night, which continues till eight or nine the next day; the hours varying a little as the days lengthen and shorten, and sometimes varying from sudden changes in the weather; as if, after being long warm, it should begin to grow cool about noon, while the air was coming down the chimney, the current will then change earlier than the usual hour, &c.

This property in chimneys I imagine we might turn to some account, and render improper, for the future, the old saying, *as useless as a chimney in summer*. If the opening of the chimney, from the breast down to the hearth, be closed by a slight moveable frame or two, in the manner of doors, covered with canvas, that will let the air through, but keep out the flies; and another little frame set within upon the hearth, with hooks on which to hang joints of meat, fowls, &c. wrapt well in wet linen cloths, three or four fold, I am confident that if the linen is kept wet, by sprinkling it once a day, the meat would be so cooled by the evaporation, carried on continually by means of the passing air, that it would keep a week or more in the hottest weather. Butter and milk might likewise be kept cool, in vessels or bottles covered with wet cloths. A shallow tray, or keeler, should be under the frame to receive any water that might drip from the wetted cloths. I think, too, that this property of chimneys might, by means of smoke-jack vanes, be applied to some mechanical purposes, where a small but pretty constant power only is wanted.

If you would have my opinion of the cause of this changing current of air in chimneys, it is, in short, as follows. In summer time there is generally a great difference in the warmth of the air at mid-day and midnight, and, of course, a difference of specific gravity in the air, as the more it is warmed the more it is rarefied. The funnel of a chimney being for the most part surrounded by the house, is protected, in a great measure, from the direction of the sun's rays, and also from the coldness of the night air. It thence preserves a middle temperature between the heat of the day, and the coldness of the night. This middle temperature it communicates to the air contained in it. If the state of the outward air be cooler than that in the funnel of the chimney, it will, by being heavier, force it to rise, and go out at the top. What supplies its place from below, being warmed, in its turn, by the warmer funnel, is likewise forced up by the colder and weightier air below, and so the current is continued till the next day, when the sun gradually changes the state of the outward air, makes it first as warm as the funnel of the chimney can make it, (when the current begins to hesitate) and afterwards warmer.

Then the funnel being cooler than the air that comes into it, cools that air, makes it heavier than the outward air, of course it descends; and what succeeds it from above, being cooled in its turn, the descending current continues till towards evening, when it again hesitates and changes its course, from the change of warmth in the outward air, and the nearly remaining same middle temperature in the funnel.

Upon this principle, if a house were built behind *Beaconhill*, an adit carried from one of the doors into the hill horizontally, till it met with a perpendicular shaft sunk from its top, it seems probable to me, that those who lived in the house, would constantly, in the heat even of the calmed day, have as much cool air passing through the house, as they should choose; and the same, though reversed in its current, during the stillest night.

I think, too, this property might be made of use to miners; as where several shafts or pits are sunk perpendicularly into the earth, communicating at bottom by horizontal passages, which is a common case, if a chimney of thirty or forty feet high were built over one of the shafts, or so near the shaft, that the chimney might communicate with the top of the shaft, all air being excluded but what should pass up or down by the shaft, a constant change of air would, by this means, be produced in the passages below, tending to secure the workmen from those damps which so frequently incommode them. For the fresh air would be almost always going down the open shaft, to go up the chimney, or down the chimney to go up the shaft. Let me add one observation more, which is, that is that part of the funnel of a chimney, which appears above the roof of a house, be pretty long, and have three of its sides exposed to the heat of the sun successively, viz. when he is in the east, in the south, and in the west, while the north side is sheltered by the building from the cool northerly winds; such a chimney will often be so heated by the sun, as to continue the draft strongly upwards, through the whole twenty four hours, and often for many days together. If the outside of such a chimney be painted black, the effect will be still greater, and the current stronger.

N° III.

IT is said the northern Chinese have a method of warming their ground floors, which is ingenious. Those floors are made of tile a foot square and two inches thick, their corners being supported by bricks set on end, that are a foot long and four inches square, the tiles, too, join into each other, by ridges and hollows along their sides. This forms a hollow under the whole floor, which on one side of the house has an opening into the air, where a fire is made, and it has a funnel rising from the other side to carry off the smoke. The fuel is a sulphurous pitcoal, the smell of which in the room is thus avoided, while the floor and of course the room is well warmed. But as the underside of the floor must grow foul with soot, and a thick coat of soot prevents much of the direct application of the hot air to the tiles, I conceive that burning the smoke by obliging it to descend through red coals, would in this construction be very advantageous, as more heat would be given by the flame than by the smoke, and the floor being thereby kept free from soot would be more heated with less fire. For this purpose I would propose erecting the funnel close to the grate, so as to have only an iron plate between the fire and the funnel, through which plate the air in the funnel being heated, it will be sure to draw well, and force the smoke to descend, as in the figure where Plate 1. A is the funnel or chimney, B the grate on which the fire is Figure 9. placed, C one of the apertures through which the descending smoke is drawn into the channel D of Figure 10, along which channel it is conveyed by a circuitous rout, as designated by the arrows, until it arrives at the small aperture E, figure 10, through which it enters the funnel F. G in both figures is the iron plate against which the fire is made, which being heated thereby, will rarefy the air in that part of the funnel, and cause the smoke to ascend rapidly. The flame thus dividing from the grate to the right and left, and turning in passages disposed, as in figure 13, so as that every part of the floor may be visited by it before it enters the funnel F, by the two passages E E, very little of the heat will be lost, and a winter room thus rendered very comfortable.

N° IV.

PAGE 8. *Few can imagine*, &c. It is said the Icelanders have very little fuel, chiefly drift wood that comes upon their coast. To receive more advantage from its heat, they make their doors low, and have a stage round the room above the door, like a gallery, wherein the women can sit and work, the men read or write, &c. The roof being tight, the warm air is confined by it and kept from rising higher and escaping; and the cold air which enters the house when the door is opened, cannot rise above the level of the top of the door, because it is heavier than the warm air above the door, and so those in the gallery are not incommoded by it. Some of our too lofty rooms might have a stage so constructed as to make a temporary gallery above, for the winter, to be taken away in summer. Sedentary people would find much comfort there in cold weather.

N° V.

PAGE 26. *Where they have the art of managing it*, &c. In some houses of the lower people among the northern nations of Europe, and among the poorer sort of Germans in Pennsylvania, I have observed this construction, which appears very advantageous. A is the kitchen with its chimney; B an iron stove in the Plate 1. stove-room. In a corner of the chimney is a hole through the Figure 11. back into the stove, to put in fuel, and another hole above it to let the smoke of the stove come back into the chimney. As soon as the cooking is over, the brands in the kitchen chimney are put through the hole to supply the stove, so that there is seldom more than one fire burning at a time. In the floor over the stove-room, is a small trap door, to let the warm air rise occasionally into the chamber. Thus the whole house is warmed at little expense of wood, and the stove-room kept constantly warm; so that in the coldest winter nights, they can work late, and find the room still comfortable when they rise to work early. An English farmer in America who makes great fires in large open chimneys, needs the constant employment of one man to cut and haul wood for supplying them; and the draft of cold air to them is so strong, that the heels of his family are frozen while they are scorching their faces, and the room is never warm, so that little sedentary work can be done by them in winter. The difference in this article alone of economy, shall, in a course of years, enable the German to buy out the Englishman, and take possession of his plantation.

MISCELLANEOUS OBSERVATIONS.

CHIMNEYS whose funnels go up in the north wall of a house and are exposed to the north winds, are not so apt to draw well as those in a south wall; because when rendered cold by those winds, they draw downwards.

Chimneys enclosed in the body of a house are better than those whose funnels are exposed in cold walls.

Chimneys in stacks are apt to draw better than separate funnels, because the funnels that have constant fires in them, warm the others in some degree that have none.

One of the funnels in a house I once occupied, had a particular funnel joined to the south side of the stack, so that three of its sides were exposed to the sun in the course of the day, viz. the east side E during the morning, the south side S in the middle Plate 1. part of the day, and the west side W during the afternoon, Figure 12. while its north side was sheltered by the stack from the cold winds. This funnel, which came from the ground floor, and had a considerable height above the roof, was constantly in a strong drawing state day and night, winter and summer.

Blacking of funnels exposed to the sun, would probably make them draw still stronger.

In Paris I saw a fire-place so ingeniously contrived as to serve conveniently two rooms, a bedchamber and a study. The funnel over the fire was round. The fire-place was of cast iron, having an upright back A, and two horizontal semicircular plates _{Plate 1.} B C, the whole so ordered as to turn on the pivots D E. The _{Figure 13.} plate B always stopped that part of the round funnel that was next to the room without fire, while the other half of the funnel over the fire was always open. By this means a servant in the morning could make a fire on the hearth C, then in the study, without disturbing the master by going into his chamber; and the master when he rose, could with a touch of his foot turn the chimney on its pivots, and bring the fire into his chamber, keep it there as long as he wanted it, and turn it again when he went out into his study. The room which had no fire in it, was also warmed by the heat coming through the back plate, and spreading in the room as it could not go up the chimney.

Plate 1

N° VI.

Description of a new STOVE for burning of Pitcoal, and consuming all its Smoke.

BY DR. B. FRANKLIN.

N° VI.

Description of a new STOVE for burning of Pitcoal, and consuming all its Smoke.

BY DR. B. FRANKLIN.

Read January 28, 1786. TOWARDS the end of the last century an ingenious French philosopher, whose name I am sorry I cannot recollect, exhibited an experiment to show that very offensive things might be burnt in the middle of a chamber, such as woollen rags, feathers, &c. without creating the least smoke or smell. The machine in which the experiment was made, if I remember right, was of this form, made of plate iron. Some clear burning charcoals were put into the opening of the short tube A, and supported there by the grate B. The air as soon as the tubes grew warm would ascend in the longer leg C and go out at D, consequently air must enter at A descending to B. In this course it must be heated by the burning coals through which it passed, and rise more forcibly in the longer tube in proportion to its degree of heat or rarefaction, and length of that tube. For such a machine is a kind of inverted syphon; and as the greater weight of water in the longer leg of a common syphon in descending is accompanied by an ascent of the same fluid in the shorter; so, in this inverted syphon, the greater quantity of levity of air in the longer leg, in rising is accompanied by the descent of air in the shorter. The things to be burned being laid on the hot coals at A, the smoke must descend through those coals, be converted into flame, which, after destroying the offensive smell, came out at the end of the longer tube as mere heated air.

Plate II. Figure I.

H   Whoever

Read January 28, 1786 TOWARDS the end of the last century an ingenious French philosopher, whose name I am sorry I cannot recollect, exhibited an experiment to show that very offensive things might be burnt in the middle of a chamber, such as woollen rags, feathers, &c. without creating the least smoke or smell. The machine in which the experiment was made, Plate II. if I remember right, was of this form, made of plate Figure I. iron. Some clear burning charcoals were put into the opening of the aloft tube A, and supported there by the grate B. The air as soon as the tubes grew warm would ascend in the longer leg C and go out at D, consequently air must enter at A descending to B. In this course it must be heated by the burning coals through which it paired, and rise more forcibly in the longer tube in proportion to its degree of heat or rarefaction, and length of that tube. For such a machine is a kind of inverted syphon; and as the greater weight of water in the longer leg of a common syphon in descending is accompanied by an ascent of the same fluid in the shorter; so, in this inverted syphon, the greater quantity of levity of air in the longer leg, in rising is accompanied by the descent of air in the shorter. The things to be burned being laid on the hot coals at A, the smoke must descend through those coals, be converted into flame, which, after destroying the offensive smell, came out at the end of the longer tube as mere heated air.

Whoever would repeat this experiment with success, must take care that the part A, B, of the short tube be quite full of burning coals, so that no part of the smoke may descend and pass by them without going through them, and being converted into flame; and that the longer tube be so heated as that the current of ascending hot air is established in it before the things to be burnt are laid on the coals; otherwise there will be a disappointment.

It does not appear either in the Memoirs of the Academy of Sciences, or Philosophical Transactions of the English Royal Society, that any improvement was ever made of this ingenious experiment, by applying it to useful purposes. But there is a German book, entitled *Vulcanus Famulans*, by Joh. George Leutmann, P. D. printed at Wirtemberg in 1723, which describes, among a great

---

Editor's Note: All figures appear together at the end of the article. See Plate 2.

variety of other stoves for warming rooms, one which seems to have been formed on the same principle, and probably from the hint thereby given, though the French experiment is not mentioned. This book being scarce, I have translated the chapter describing the stove, viz.

"Vulcanus Famulans, by John George Leutmann, P. D. Wirtemberg, 1723.

C H A P. VII.

"On a stove, which draws downwards.

"Here follows the description of a sort of stove, which can easily be removed and again replaced at pleasure. This drives the fire down under itself, and gives no smoke, but however a very unwholesome vapor.

Plate II.   "In the figure, A is an iron vessel like a funnel, in diameter
Figure 20. at the top about twelve inches, at the bottom near the grate about five inches; its height twelve inches. This is set on the barrel C, which is ten inches diameter and two feet long, closed at each end E E. From one end rises a pipe or flue about four inches diameter, on which other pieces of pipe are set, which are gradually contracted to D, where the opening is but about two inches. Those pipes must together be at least four feet high. B is an iron grate. F F are iron handles guarded with wood, by which the stove is to be lifted and moved. It stands on three legs. Care must be taken to stop well all the joints, that no smoke may leak through.

"When this stove is to be used, it must first be carried into the kitchen and placed in the chimney near the fire. There burning wood must be laid and left upon its grate till the barrel C is warm, and the smoke no longer rises at A, but descends towards C. Then it is to be carried into the room which it is to warm. When once the barrel C is warm, fresh wood may be thrown into the vessel A as often as one pleases, the flame descends and without smoke, which is so consumed that only a vapor passes out at D.

"As this vapor is unwholesome, and affects the head, one may be freed from it, by fixing in the wall of the room an inverted funnel, such as people use to hang over lamps, through which their smoke goes out as through a chimney. This funnel carries out all the vapor cleverly, so that one finds no inconvenience from it, even though the opening D be placed a span below the mouth of the said funnel G. The neck of the funnel is better when made gradually bending, than if turned in a right angle.

"The cause of the draft downwards in the stove is the pressure of the outward air, which falling into the vessel A in a column of twelve inches diameter, finds only a resisting passage at the grate B, of five inches, and one at D, of two inches, which are much too weak to drive it back again; besides, A stands much higher than B, and so the pressure on it is greater and more forcible, and beats down the flame to that part where it finds the least resistance. Carrying the machine first to the kitchen fire for preparation, is on this account, that in the beginning the fire and smoke naturally ascend, till the air in the close barrel C is made thinner by the warmth. When that vessel is heated, the air in it is rarefied, and then all the smoke and fire descends under it.

"The wood should be thoroughly dry, and cut into pieces five or six inches long, to fit it for being thrown into the funnel A." Thus far the German book.

It appears to me by Mr. Leutmann's explanation of the operation of this machine, that he did not understand the principles of it, whence I conclude he was not the inventor of it; and by the description of it, wherein the opening at A is made so large, and the pipe E, D, so short, I am persuaded he never made nor saw the experiment, for the first ought to be much smaller and the last much higher, or it hardly will succeed. The carrying it in the kitchen, too, every time the fire should happen to be out, must be so troublesome, that it is not likely ever to have been in practice, and probably has never been shown but as a philosophical experiment. The funnel for conveying the vapor out of the room, would besides have been uncertain in its operation, as a wind blowing against its mouth would drive the vapor back.

The stove I am about to describe, was also formed on the idea given by the French experiment, and completely carried into execution before I had any knowledge of the German invention;

which I wonder should remain so many years in a country where men are so ingenious in the management of fire, without receiving long since the improvements I have given it.

DESCRIPTION

DESCRIPTION of the PARTS.

A, the bottom plate which lies flat upon the hearth, with its partitions I, 2, 3, 4, 5, 6, that are cast with it, and a groove Z Z, in which are to slide, the bottom edges of the small plates Y, Y, figure 2; which plates meeting at X close the front. Plate II.
Figure 2.

B 1, figure 3, is the cover plate showing its under side, with the grooves 1, 2, 3, 4, 5, 6, to receive the top edges of the partitions that are fixed to the bottom plate. It shows also the grate W W, the bars of which are cast in the plate, and a groove V V, which comes right over the groove Z Z, figure 2, receiving the upper edges of the small sliding plates Y Y, figure 12.

B 2, figure 4, shows the upper side of the same plate, with a square impression or groove for receiving the bottom mouldings T T T T of the three sided box C, figure 5, which is cast in one piece.

D, figure 6, its cover, showing its under side with grooves to receive the upper edges S S S of the sides of C, figure 5, also a groove R, R, which when the cover is put on comes right over another QQ in C, figure 5, between which it is to slide.

E, figure 7, the front plate of the box.

P, a hole three inches diameter through the cover D, figure 6, over which hole stands the vase F, figure 8, which has a corresponding hole two inches diameter through its bottom.

The top of the vase opens at O, O, O, figure 8, and turns back upon a hinge behind when coals are to be put in; the vase has a grate within at N N of cast iron H, figure 9, and a hole in the top one and a half inches diameter to admit air, and to receive the ornamental brass guilt flame M, figure 10, which stands in that hole and, being itself hollow and open, suffers air to pass through it to the fire.

G, figure 11, is a drawer of plate iron, that slips in between in the partitions 2 and 3, figure 2, to receive the falling ashes. It is concealed when the small sliding plates Y Y, figure 12, are shut together.

I, I, I, I, figure 8, is a niche built of brick in the chimney and plastered. It closes the chimney over the vase, but leaves two funnels one in each corner communicating with the bottom box K K, figure 2.

DIMENSIONS of the PARTS.

	Feet.	In.
Front of the bottom box,	2	0
Height of its partitions,	0	$4^1/4$
Length of N° 1, 2, 3 and 4, each,	1	3
Length of N° 5 and 6, each	0	$8^1/4$
Breadth of the passage between N° 2 and 3,	0	6
Breadth of the other passages each,	0	$3^1/2$
Breadth of the grate,	0	$6^1/2$
Length of ditto,	0	8
Bottom moulding of box C, square,	1	0
Height of the sides of ditto,	0	4
Length of the back side,	0	10
Length of the right and left sides, each,	0	$9^1/2$
Length of the front plate E, where longest,	0	11
The cover D, square,	0	12
Hole in ditto, diameter,	0	3
Sliding plates Y Y their length, each,	1	0
———— their breadth, each,	0	$4^1/2$
Drawer G, its length,	1	0
———— breadth,	0	$5^1/4$
———— depth,	0	4
———— depth of its further end, only,	0	1
Grate H in the vase, its diameter to the extremity of its knobs,	0	$5^3/4$
Thickness of the bars at top,	0	$0^1/4$
———— at bottom, less,	0	0
Depth of the bars at the top,	0	$0^3/4$

Height of the vase,	- - - 1	6
Diameter of the opening O, O, in the clear,	0	8
Diameter of the air-hole at top,	- 0	1½
——————— of the flame hole at bottom,	- 0	2

To fix this Machine.

Spread mortar on the hearth to bed the bottom plate A, then lay that plate, level, equally distant from each jamb, and projecting out as far as you think proper. Then putting some Windsor loam in the grooves of the cover B, lay that on: Trying the sliding plates Y Y, to see if they move freely in the groves Z Z, V V, designed for them.

Then begin to build the niche, observing to leave the square corners of the chimney unfilled; for they are to be funnels. And observe also to leave a free open communication between the passages at K K, and the bottom of those funnels, and mind to close the chimney above the top of the niche, that no air may pass up that way. The concave back of the niche will rest on the circular iron partition I A 4, figure 2, then with a little loam put on the box C over the grate, the open side of the box in front.

Then, with loam in three of its grooves, the groove R R being left clean, and brought directly over the groove Q Q in the box, put on the cover D, trying the front plate E, to see if it slides freely in those grooves.

Lastly, set on the vase, which has small holes in the moulding of its bottom to receive two iron pins that rise out of the plate D at I I, for the better keeping it steady.

Then putting in the grate H, which rests on its three knobs H H H against the inside of the vase, and slipping the drawer into its place; the machine is fit for use.

To use it.

Let the first fire be made after eight in the evening or before eight in the morning, for at those times and between those hours all night, there is usually a draft up a chimney, though it has long been without fire; but between those hours in the day there is often in a cold chimney a draft downwards, when if you attempt to kindle a fire, the smoke will come into the room.

But to be certain of your proper time, hold a flame over the air-hole at the top. If the flame is drawn strongly down for a continuance, without whiffling, you may begin to kindle a fire.

First put in a few charcoals on the grate H.

Lay some small sticks on the charcoals,

Lay some pieces of paper on the sticks,

Kindle the paper with a candle,

Then shut down the top, and the air will pass down through the air-hole, blow the flame of the paper down through the sticks, kindle them, and their flame passing lower, kindles the charcoal.

When the charcoal is well kindled, lay on it the sea-coals, observing not to choke the fire by putting on too much at first.

The flame descending through the hole in the bottom of the vase, and that in plate D into the box C passes down farther through the grate W W in plate B I, then passes horizontally towards the back of the chimney; there dividing, and turning to the right and left, one part of it passes round the far end of the partition 2, then coming forward it turns round the near end of partition 2, then moving backward it arrives at the opening into the bottom of one of the upright corner funnels behind the niche, through which it ascends into the chimney, thus heating that half of the box and that side of the niche. The other part of the divided flame passes round the far end of partition 3, round the near end of partition 4, and so into and up the other corner funnel, thus heating the other half of the box, and the other side of the niche. The vase itself, and the box C will also be very hot, and the air surrounding them being heated, and rising, as it cannot get into the chimney, it spreads in the room, colder air succeeding is warmed in its turn, rises and spreads, till by the continual circulation the whole is warmed.

If you should have occasion to make your first fire at hours not so convenient as those above mentioned, and when the chimney does not draw, do not begin it in the vase, but in one or more of the passages of the lower plates first covering the mouth of the vase. After the chimney has drawn a while with the fire thus low, and begins to be a little warm, you may close those passages and kindle another fire in the box C, leaving its sliding shutter a little open; and when you find after some time that the chimney being warmed draws forcibly, you may shut that passage, open your vase, and kindle your fire there, as above directed. The chimney well warmed by the first day's fire will continue to draw constantly all winter, if fires are made daily.

You will, in the management of your fire, have need of the following implements:

A pair of small light tongs, twelve or fifteen inches long, plate II, figure 13.

A light poker about the same length with a flat broad point, figure 14.

A rake to draw allies out of the passages of the lower plate, where the lighter kind escaping the ash-box will gather by degrees, and perhaps once in a week or ten days require being removed, figure 15.

And a fork with its prongs wide enough to flip on the neck of the vase cover, in order to raise and open it when hot, to put in fresh coals, figure 16.

In the management of this stove there are certain precautions to be observed, at first with attention, till they become habitual. To avoid the inconvenience of smoke, see that the grate H be clear before you begin to light a fresh fire. If you find it clogged with cinders and ashes, turn it up with your tongs and let them fall upon the grate below the ashes will go through it, and the cinders may be raked off and returned into the vase when you would burn them. Then see that all the sliding plates are in their places and close shut, that no air may enter the stove but through the round opening at the top of the vase. And to avoid the inconvenience of dust from the ashes, let the ash-drawer be taken out of the room to be emptied; and when you rake the passages, do it when the draft of the air is strong inwards, and put the ashes carefully into the ash-box, that remaining in its place.

If being about to go abroad, you would prevent your fire burning in your absence, you may do it by taking the brass flame from the top of the vase, and covering the passage with a round tin plate, which will prevent the entry of more air than barely sufficient to keep a few of the coals alive. When you return, though some hours absent, by taking off the tin plate and admitting the air, your fire will soon be recovered.

The effect of this machine, well managed, is to burn not only the coals, but all the smoke of the coals, so that while the fire is burning, if you go out and observe the top of your chimney, you will see no smoke issuing, nor any thing but clear warm air, which as usual makes the bodies seen through it appear waving.

But let none imagine from this, that it may be a cure for bad or smoky chimneys, much less, that as it burns the smoke it may be used in a room that has no chimney. 'Tis by the help of a good chimney, the higher the better, that it produces its effect; and though a flue of plate iron sufficiently high might be raised in a very lofty room, the management to prevent all disagreeable vapor would be too nice for common practice, and small errors would have unpleasing consequences.

It is certain that clean iron yields no offensive smell when heated. Whatever of that kind you perceive, where there are iron stoves, proceeds therefore from some foulness burning or fuming on their surface. They should therefore never be spit upon, or greased, nor should any dust be suffered to lie upon them. But as the greatest care will not always prevent these things, it is well once a week to wash the stove with soap lees and a brush, rinsing it with clean water.

The Advantages of this Stove.

1. The chimney does not grow foul, nor ever need sweeping; for as no smoke enters it, no soot can form in it.

2. The air heated over common fires instantly quits the room and goes up the chimney with the smoke; but in the stove; it is obliged to descend in flame and pass through the long winding horizontal passages, communicating its heat to a body of iron plate, which having thus time to receive the heat, communicates the same to the air of the room, and thereby warms it to a greater degree.

3. The whole of the fuel is consumed by being turned into flame, and you have the benefit of its heat, whereas in common

chimneys a great part goes away in smoke which you see as it rises, but it affords you no rays of warmth. One may obtain some notion of the quantity of fuel thus wasted in smoke, by reflecting on the quantity of soot that a few weeks firing will lodge against the sides of the chimney, and yet this is formed only of those particles of the column of smoke that happen to touch the sides in its ascent. How much more must have passed off in the air? And we know that this soot is still fuel; for it will burn and flame as such, and when hard caked together is indeed very like and almost as solid as the coal it proceeds from. The destruction of your fuel goes on nearly in the same quantity whether in smoke or in flame: but there is no comparison in the difference of heat given. Observe when fresh coals are first put on your fire, what a body of smoke arises. This smoke is for a long time too cold to take flame. If you then plunge a burning candle into it, the candle instead of inflaming the smoke will instantly be itself extinguished. Smoke must have a certain degree of heat to be inflammable. As soon as it has acquired that degree, the approach of a candle will inflame the whole body, and you will be very sensible of the difference of the heat it gives. A still easier experiment may be made with the candle itself. Hold your hand near the side of its flame, and observe the heat it gives; then blow it out, the hand remaining in the same place, and observe what heat may be given by the smoke that rises from the still burning snuff. You will find it very little. And yet that smoke has in it the substance of so much flame, and will instantly produce it, if you hold another candle above it so as to kindle it. Now the smoke from the fresh coals laid on this stove, instead of ascending and leaving the fire while too cold to burn, being obliged to descend through the burning coals, receives among them that degree of heat which converts it into flame, and the heat of that flame is communicated to the air of the room, as above explained.

4. The flame from the fresh coals laid on in this stove, descending through the coals already ignited, preserves them long from consuming, and continues them in the state of red coals as long as the flame continues that surrounds them, by which means the fires made in this stove are of much longer duration than in any other, and fewer coals are therefore necessary for a day. This

is a very material advantage indeed. That flame should be a kind of pickle, to preserve burning coals from consuming, may seem a paradox to many, and very unlikely to be true, as it appeared to me the first time I observed the fact. I must therefore relate the circumstances, and shall mention an easy experiment, by which my reader may be in possession of every thing necessary to the understanding of it. In the first trial I made of this kind of stove, which was constructed of thin plate iron, I had instead of the vase a kind of inverted pyramid like a mill-hopper; and fearing at first that the small grate contained in it might be clogged by cinders, and the passage of the flame sometimes obstructed, I ordered a little door near the grate, by means of which I might on occasion clear it. Though after the stove was made, and before I tried it, I began to think this precaution superfluous, from an imagination, that the flame being contracted in the narrow part where the grate was placed, would be more powerful in consuming what it should there meet with, and that any cinders between or near the bars would be presently destroyed and the passage opened. After the stove was fixed and in action, I had a pleasure now and then in opening that door a little, to see through the crevice how the flame descended among the red coals, and observing once a single coal lodged on the bars in the middle of the focus, a fancy took me to observe by my watch in how short a time it would be consumed. I looked at it long without perceiving it to be at all diminished, which surprised me greatly. At length it occurred to me, that I and many others had seen the same thing thousands of times, in the conservation of the red coal formed in the snuff of a burning candle, which while enveloped in flame, and thereby prevented from the contact of passing air, is long continued and augments instead of diminishing, so that we are often obliged to remove it by the snuffers, or bend it out of the flame into the air, where it consumes presently to ashes. I then supposed that to consume a body by fire, passing air was necessary to receive and carry off the separated particles of the body; and that the air passing in the flame of my stove, and in the flame of a candle, being already saturated with such particles, could not receive more, and therefore left the coal undiminished as long as the outward air was prevented from coming to it by the surrounding flame, which kept it in a situation somewhat like that of charcoal in a well luted crucible, which, though long kept in a strong fire, comes out unconsumed.

An easy experiment will satisfy any one of this conserving power of flame enveloping red coal. Take a small stick of deal or other wood the size of a goose quill, and hold it horizontally and steadily in the flame of the candle above the wick, without touching it, but in the body of the flame. The wood will first be inflamed, and burn beyond the edge of the flame of the candle, perhaps a quarter of an inch. When the flame of the wood goes out, it will leave a red coal at the end of the stick, part of which will be in the flame of the candle and part out in the air. In a minute or two you will perceive the coal in the air diminish gradually, so as to form a neck; while the part in the flame continues of its first size, and at length the neck being quite consumed it drops off; and by rolling it between your fingers when extinguished you will find it still a solid coal.

However, as one cannot be always putting on fresh fuel in this stove to furnish a continual flame as is done in a candle, the air in the intervals of time gets at the red coals and consumes them. Yet the conservation while it lasted, so much delayed the consumption of the coals, that two fires, one made in the morning, and the other in the afternoon, each made by only a hatful of coals, were sufficient to keep my writing room, about sixteen feet square and ten high, warm a whole day. The fire kindled at seven in the morning would burn till noon; and all the iron of the machine with the walls of the niche being thereby heated, the room kept warm till evening, when another smaller fire kindled kept it warm till midnight.

Instead of the sliding plate E, which shuts the front of the box C, I sometimes used another which had a pane of glass, or, which is better, of Muscovy talc, that the flame might be seen descending from the bottom of the vase and passing in a column through the box C, into the cavities of the bottom plate, like water falling from a funnel, admirable to such as are not acquainted with the nature of the machine, and in itself a pleasing spectacle.

Every utensil, however properly contrived to serve its purpose, requires some practice before it can be used adroitly. Put into the hands of a man for the first time, a gimblet or a hammer, (very simple instruments) and tell him the use of them, he shall neither bore a hole or drive a nail with the dexterity or success of another who has been a little accustomed to handle them. The beginner therefore in the use of this machine, will do well not to be discouraged with little accidents that may arise at first from his want of experience. Being somewhat complex, it requires as already said a variety of attentions; habit will render them necessary. And the studious man who is much in his chamber, and has a pleasure in managing his own fire, will soon find this a machine most comfortable and delightful. To others who leave their fires to the care of ignorant servants, I do not recommend it. They will with difficulty acquire the knowledge necessary, and will make frequent blunders that will fill your room with smoke. It is therefore by no means fit for common use in families. It may be advisable to begin with the flaming kind of stone coal, which is large, and, not caking together, is not so apt to clog the grate. After some experience, any kind of coal may be used, and with this advantage, that no smell, even from the most sulfurous kind can come into your room, the current of air being constantly into the vase, where too that smell is all consumed.

The vase form was chosen as being elegant in itself, and very proper for burning of coals: Where wood is the usual fuel, and must be burnt in pieces of some length, a long square   Plate 2. chest may be substituted, in which A is the cover open-   Figure 17. ing by a hinge behind, B the grate, C the hearth box with its divisions as in the other, D the plan of the chest, E the long narrow grate. This I have not tried, but the vase machine was completed in 1771, and used by me in London three winters, and one afterwards in America, much to my satisfaction; and I have not yet thought of any improvement it may be capable of, though such may occur to others. For common use, while in France, I have contrived another grate for coals, which has in part the same property of burning the smoke and preserving the red coals longer by the flame, though not so completely, as in the vase, yet sufficiently to be very useful, which I shall now describe as follows.

A, is a round grate, one foot (French) in diameter,   Plate 2. and eight inches deep between the bars and the back;   Figure 18. the sides and back of plate iron; the sides having holes of half an inch diameter distant 3 or 4 inches from each other, to let in air for enlivening the fire. The back without holes. The sides do not meet at top nor at bottom by eight inches: that square is filled by grates of small bars crossing front to back to let in air below, and let out the smoke or flame above. The three middle bars of the front grate are fixed, the upper and lower may be taken out and put in at pleasure, when hot, with a pair of pincers. This round grate turns upon an axis, supported by the crotchet B, the stem of which is an inverted conical tube five inches deep, which comes on as many inches upon a pin that fits it, and which is fixed upright in a cast iron plate D, that lies upon the hearth; in the middle of the top and bottom grates are fixed small upright pieces E E about an inch high, which as the whole is turned on its axis stop it when the grate is perpendicular. Figure 19 is another view of the same machine.

In making the first fire in a morning with this grate, there is nothing particular to be observed. It is made as in other grates, the coals being put in above, after taking out the upper bar, and replacing it when they are in. The round figure of the fire when thoroughly kindled is agreeable, it represents the great giver of warmth to our system. As it burns down and leaves a vacancy above, which you would fill with fresh coals, the upper bar is to be taken out, and afterwards replaced. The fresh coals while the grate continues in the same position, will throw up as usual a body of thick smoke. But every one accustomed to coal fires in common grates, must have observed that pieces of fresh coal stuck in below among the red coals have their smoke so heated as that it becomes flame as fast as it is produced, which flame rises among the coals and enlivens the appearance of the fire. Here then is the use of this swivel grate. By a push with your tongs or poker, you turn it on its pin till it faces the back of the chimney, then turn it over on its axis gently till it again faces the room, whereby all the fresh coals will be found under the live coals, and the greater part of the smoke arising from the fresh coals will in its passage through the live ones be heated so as to be converted into flame: Whence you have much more heat from them, and your red coals are longer preserved from consuming. I conceive this construction, though not so complete a consumer of all the smoke as the vase, yet to be fitter for common use, and very advantageous. It gives too a full sight of the fire, always a pleasing object, which we have not in the other. It may with a touch be turned more or less from any one of the company that desires to have less of its heat, or presented full to one just come out of the cold. And supported in a horizontal position, a tea-kettle may be boiled on it.

The author's description of his Pennsylvania fire-place, first published in 1744, having fallen into the hands of workmen in Europe, who did not, it seems, well comprehend the principles of that machine, it was much disfigured in their imitations of it; and one of its main intentions, that of admitting a sufficient quantity of fresh air warmed in entering through the air-box, nearly defeated, by a pretended improvement, in lessening its passages to make more room for coals in a grate. On pretense of such improvements, they obtained patents for the invention, and for a while made great profit by the sale, till the public became sensible of that defect, in the expected operation. If the same thing should be attempted with this vase stove, it will be well for the buyer to examine thoroughly such pretended improvements, left, being the mere productions of ignorance, they diminish or defeat the advantages of the machine, and produce inconvenience and disappointment.

The method of burning smoke, by obliging it to descend through hot coals, may be of great use in heating the walls of a hot-house. In the common way, the horizontal passages or flues that are made to go and return in those walls, lose a great deal of their effect when they come to be foul with soot; for a thick blanket-like lining of soot prevents much of the hot air from touching and heating the brick work in its passage, so that more fire must be made as the flue grows fouler: But by burning the smoke they are kept always clean. The same method may also be of great advantage to those businesses in which large coppers or caldrons are to be heated.

Written at Sea, 1785.

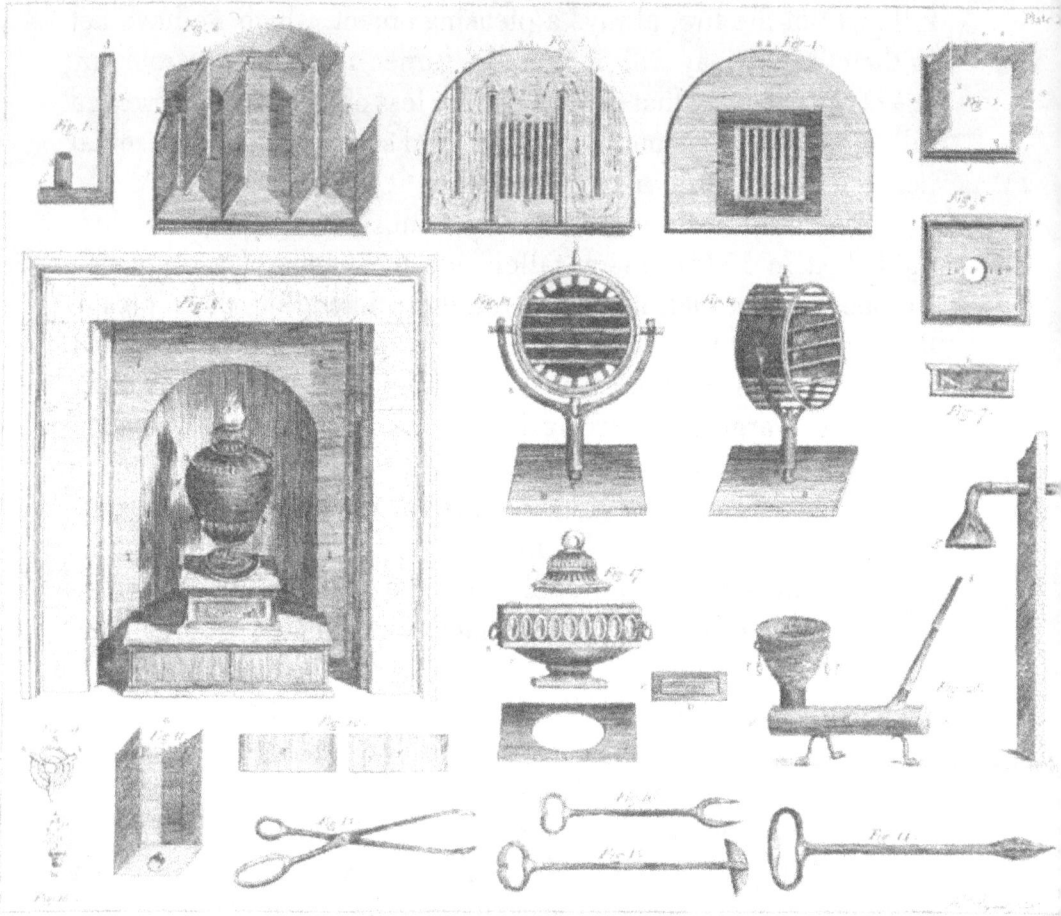

Nº XXXVIII.

A Letter from Dr. BENJAMIN FRANKLIN, to Mr. ALPHONSUS le ROY, *Member of Several Academies, at Paris. Containing Sundry Maritime Observations.*

N° XXXVIII.

A Letter from Dr. BENJAMIN FRANKLIN, *to* Mr. AL-PHONSUS le Roy, *Member of several Academies, at Paris. Containing sundry Maritime Observations.*

At Sea, on board the London Packet; Capt. Truxton, August 1785.

SIR,

Read Dec. 2, 1785.

YOUR learned writings on the navigation of the antients, which contain a great deal of curious information; and your very ingenious contrivances for improving the modern sails (*voilure*) of which I saw with great pleasure a successful trial on the river Seine, have induced me to submit to your consideration and judgment, some thoughts I have had on the latter subject.

Those mathematicians who have endeavoured to improve the swiftness of vessels, by calculating to find the form of least resistance, seem to have considered a ship as a body moving through one fluid only, the water; and to have given little attention to the circumstance of her moving through another fluid, the air. It is true that when a vessel sails right before the wind, this circumstance is of no importance, because the wind goes with her; but in every deviation from that course, the resistance of the air is something, and becomes greater in proportion as that deviation increases. I wave at present the consideration of those different degrees of resistance given by the air to that part of the hull which is above water, and confine myself to that given to the sails; for their motion through the air is resisted by the air, as the motion of the hull through the water is resisted by the water, though with less force as the air is a lighter fluid. And to simplify the discussion as much as possible, I would state one situation only, to wit, that of the wind upon the beam, the ship's course being directly across the wind; and I would

suppose

At Sea, on board the London Packet, Capt. Truxton, August 1785.

SIR,

Read Dec.
2, 1875. YOUR learned writings on the navigation of the
ancients, which contain a great deal of curious information; and your very ingenious contrivances for improving the
modern sails (*voilure*) of which I saw with great pleasure a successful trial on the river Seine, have induced me to submit to your
consideration and judgment, some thoughts I have had on the
latter subject.

Those mathematicians who have endeavoured to improve the
swiftness of vessels, by calculating to find the form of least resistance, seem to have considered a ship as a body moving through
one fluid only, the water; and to have given little attention to the
circumstance of her moving through another fluid, the air. It is true
that when a vessel sails right before the wind, this circumstance is
of no importance, because the wind goes with her; but in every
deviation from that course, the resistance of the air is something,
and becomes greater in proportion as that deviation increases. I
wave at present the consideration of those different degrees of
resistance given by the air to that part of the hull which is above
water, and confine myself to that given to the sails; for their motion
through the air is resisted by the air, as the motion of the hull
through the water is resisted by the water, though with less force
as the air is a lighter fluid. And to simplify the discussion as much
as possible, I would state one situation only, to wit, that of the
wind upon the beam, the ship's course being directly across the
wind; and I would suppose the sail set in an angle of 45 degrees
with the keel, as in the following figure; wherein A B represents
the body of the vessel, CD the position of the sail, EEE the direction
of the wind, MM the line of motion. In observing this figure it will
appear, that so much of the body of the vessel as is immersed in
the water, must, to go forward, remove out of its way what water
it meets with between the pricked lines FF. And the sail, to go
forward, must move out of its way all the air its whole dimension

Editor's note: All figures appear in Plate 4.

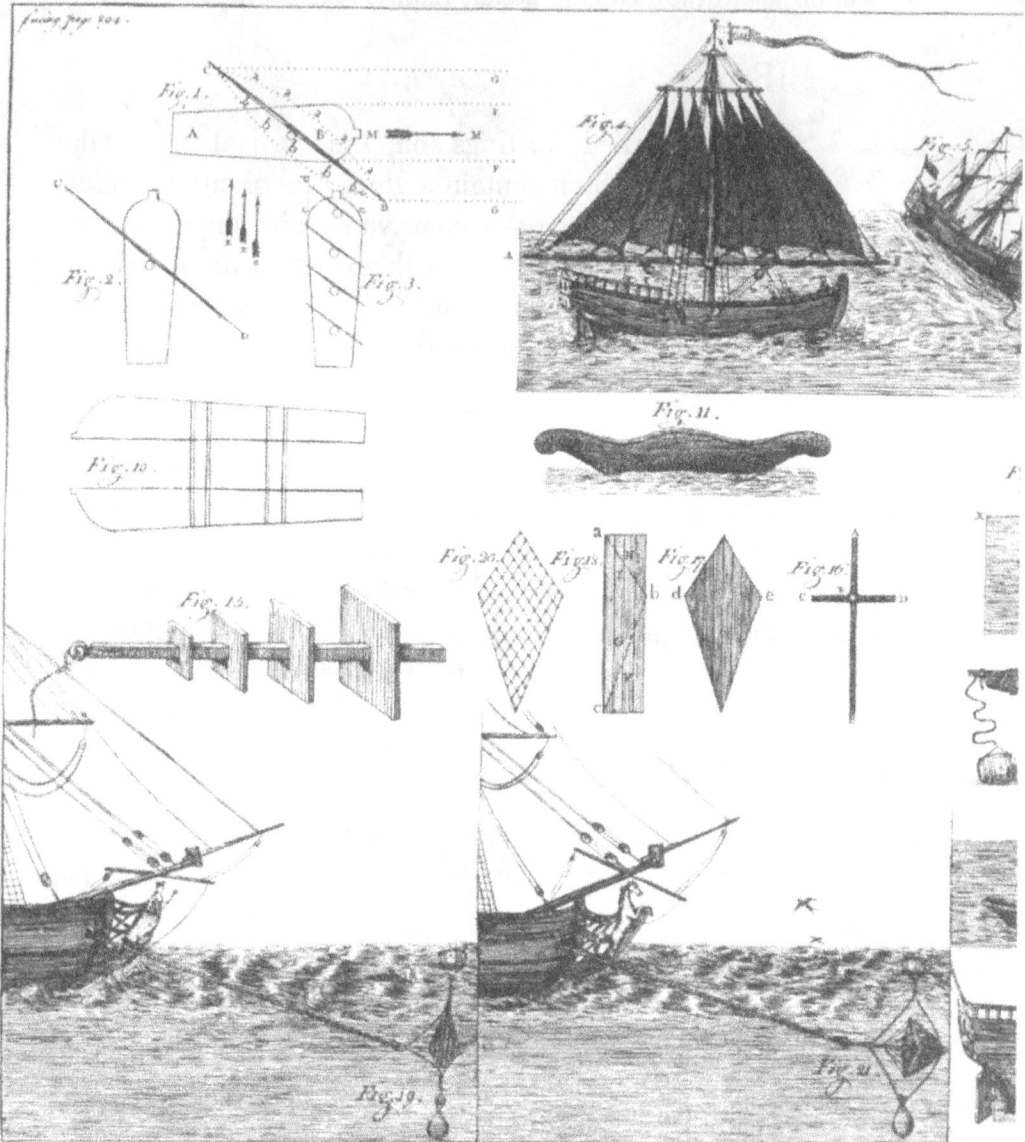

meets with between the pricked lines CG and DG. Thus both the fluids give resistance to the motion, each in proportion to the quantity of matter contained in the dimension to be removed. And though the air is vastly lighter than the water, and therefore more easily removed, yet the dimension being much greater its effect is very considerable.

It is true that in the case stated, the resistance given by the air between those lines to the motion of the sail is not apparent to the eye, because the greater force of the wind which strikes it

in the direction E E E, overpowers its effect, and keeps the sail full in the curve a, a, a, a, a. But suppose the wind to cease, and the vessel in a calm to be impelled with the same swiftness by oars, the sail would then appear filled in the contrary curve b, b, b, b, b, when prudent men would immediately perceive that the air resisted its motion, and would order it to be taken in.

Is there any possible means of diminishing this resistance, while the same quantity of sail is exposed to the action of the wind, and therefore the same force obtained from it? I think there is, and that it may be done by dividing the sail into a number of parts, and placing those parts in a line one behind the other; thus instead of one sail extending from C to D, figure 2, if four sails containing together the same quantity of canvas, were placed as in figure 3, each having one quarter of the dimensions of the great sail, and exposing a quarter of its surface to the wind, would give a quarter of the force; so that the whole force obtained from the wind would be the same, while the refinance from the air would be nearly reduced to the space between the pricked lines *ab* and *cd*, before the foremost sail.

It may perhaps be doubted whether the resistance from the air would be so diminished; since possibly each of the following small sails having also air before it, which must be removed, the refinance on the whole would be the same.

This is then a matter to be determined by experiment. I will mention one that I many years since made with success for another purpose; and I will propose another small one easily made. If that too succeeds, I should think it worth while to make a larger, though at some expense, on a river boat; and perhaps time and the improvements experience will afford, may make it applicable with advantage to larger vessels.

Having near my kitchen chimney a round hole of eight inches diameter, through which was a constant steady current of air, increasing or diminishing only as the fire increased or diminished, I contrived to place my jack so as to receive that current; and taking off the flyers, I fixed in their stead on the same pivot a round tin plate of near the same diameter with the hole; and having cut it in radial lines almost to the centre, so as to have six equal vanes, I gave to each of them the obliquity of forty-five degrees. They moved round, without the weight, by the impression only of the current of air, but too slowly for the purpose of roasting. I suspected that the air struck by the back of each vane might possibly by its resistance retard the motion; and to try this, I cut each of them into two, and I placed the twelve, each having the same obliquity, in a line behind each other, when I perceived a great augmentation in its velocity, which encouraged me to divide them once more, and, continuing the same obliquity, I placed the twenty-four behind each other in a line, when the force of the wind being the same, and the surface of vane the same, they moved round with much greater rapidity, and perfectly answered my purpose.

The second experiment that I propose, is, to take two playing cards of the same dimensions, and cut one of them transversely into eight equal pieces; then with a needle string them upon two threads one near each end, and place them so upon the threads that, when hung up, they may be one exactly over the other, at a distance equal to their breadth, each in a horizontal position; and let a small weight, such as a bird-shot, be hung under them, to make them fall in a straight line when let loose. Suspend also the whole card by threads from its four corners, and hang to it an equal weight, so as to draw it downwards when let fall, its whole breadth pressing against the air. Let those two bodies be attached, one of them to one end of a thread a yard long, the other to the other end. Extend a twine under the ceiling of a room, and put through it at thirty inches distance two pins bent in the form of fishhooks. On these two hooks hang the two bodies, the thread that connects them extending parallel to the twine, which thread being cut, they must begin to fall at the same instant. If they take equal time in falling to the floor, it is a proof that the resistance of the air is in both cases equal. If the whole card requires a longer time, it shows that the sum of the resistances to the pieces of the cut card is not equal to the resistance of the whole one.*

This principle so far confirmed, I would proceed to make Plate IV. a larger experiment, with a shallop, which I would rig in Figure 4. this manner.

AB is a long boom, from which are hoisted seven jibs, a, b, c, d, e, f, g, each a seventh part of the whole dimensions, and as much more as will fill the whole space when set in an angle of forty-five degrees, so that they may lap when going before the wind, and hold more wind when going large. Thus rigged, when going right before the wind, the boom should be brought at right angles with the keel, by means of the sheet ropes C D, and all the sails hauled flat to the boom.

---

*The motion of the vessel made it inconvenient to try this simple experiment, at sea, when the proposal of it was written. But it has been tried since we came on shore, and succeeded as the other.

These positions of boom and sails to be varied as the wind quarters. But when the wind is on the beam, or when you would turn to windward, the boom is to be hauled right fore and aft, and the sails trimmed according as the wind is more or less against your course.

It seems to me that the management of a shallop so rigged would be very easy, the sails being run up and down separately, so that more or less sail may be made at pleasure; and I imagine, that there being full as much sail exposed to the force of the wind which impells the vessel in its course, as if the whole were in one piece, and the resistance of the dead air against the foreside of the sail being diminished, the advantage of swiftness would be very considerable; besides that the vessel would lie nearer the wind.

Since we are on the subject of improvements in navigation, permit me to detain you a little longer with a small relative observation. Being, in one of my voyages, with ten merchant-ships under convoy of a frigate at anchor in Torbay, waiting for a wind to go to the westward; it came fair, but brought in with it a considerable swell. A signal was given for weighing, and we put to sea all together; but three of the ships left their anchors, their cables parting just as the anchors came a-peak. Our cable held, and we got up our anchor; but the shocks the ship felt before the anchor got loose from the ground, made me reflect on what might possibly have caused the breaking of the other cables; and I imagined it might be the short bending of the cable just without the hause-hole, from a horizontal to an almost vertical position, and the sudden violent jerk it receives by the rising of the head of the ship on the swell of a wave while in that position. For example, suppose a vessel hove up so as to have her head nearly over her anchor, which still keeps its hold, perhaps in a tough bottom; if it were calm, the cable still out would form nearly a perpendicular line, measuring the distance between the hause-hole [hawsehole] and the anchor; but if there is a swell, her head in the trough of the sea will fall below the level, and when lifted on the wave will be as much above it. In the first case the cable will hang loose and bend perhaps as in figure 5. In the second case figure 6, the cable will be drawn straight with a jerk, must sustain the whole force of the riling ship, and must either loosen the anchor, resist the rising force of the ship, or break. But why does it break at the hause-hole?

Let us suppose it a cable of three inches diameter, and represented by figure 7. If this cable is to be bent round the corner A, it is evident that either the part of the triangle contained between the letters a, b, c, must stretch considerably, and those most that are nearest the surface; or that the parts between d, e, f, must be compressed; or both, which most probably happens. In this case the lower half of the thickness affords no strength against the jerk, it not being strained, the upper half bears the whole, and the yarns near the upper surface being first and most strained, break first, and the next yarns follow; for in this bent situation they cannot bear the strain all together, and each contribute its strength to the whole, as they do when the cable is strained in a straight line.

To remedy this, methinks it would be well to have a kind of large pulley wheel, fixed in the hause-hole, suppose of two feet diameter, over which the cable might pass; and being there bent gradually to the round of the wheel, would thereby be more equally strained, and better able to bear the jerk, which may save the anchor, and by that means in the course of the voyage may happen to save the ship.

One maritime observation more shall finish this letter. I have been a reader of news-papers now near seventy years, and I think few years pass without an account of some vessel met with at sea, with no soul living on board, and so many feet of water in her hold, which vessel has nevertheless been saved and brought into port: and when not met with at sea, such forsaken vessels have often come ashore on some coast. The crews who have taken to their boats and thus abandoned such vessels, are sometimes met with and taken up at sea by other ships, sometimes reach a coast, and are sometimes never heard of. Those that give an account of quitting their vessels, generally say, that she sprung a leak, that they pumped for some time, that the water continued to rise upon them, and that despairing to save her, they had quitted her lest they should go down with her. It seems by the event that this fear was not always well founded, and I have endeavoured to guess at the reason of the people's too hasty discouragement.

When a vessel springs a leak near her bottom, the water enters with all the force given by the weight of the column of water, without, which force is in proportion to the difference of level between the water without and that within. It enters therefore with more force at first, and in greater quantity, than it can afterwards when the water within is higher. The bottom of the vessel too is narrower, so that the same quantity of water coming into that narrow part, rises faster than when the space for it to flow in is larger. This helps to terrify. But as the quantity entering is less and less as the surfaces without and within become more nearly equal in height, the pumps that could not keep the water from rising at first, might afterwards be able to prevent its rising higher, and the people might have remained on board in safety, without hazarding themselves in an open boat on the wide ocean. (Fig. 8.)

Besides the greater equality in the height of the two surfaces, there may sometimes be other causes that retard the farther sinking of a leaky vessel. The rising water within may arrive at quantities of light wooden work, empty chests, and particularly empty water casks, which if fixed so as not to float themselves may help to sustain her. Many bodies which compose a ship's cargo may be specifically lighter than water, all these when out of water are an additional weight to that of the ship, and she is in proportion pressed deeper into the water; but as soon as these bodies are immersed, they weigh no longer on the ship, but on the contrary, if fixed, they help to support her, in proportion as they are specifically lighter than the water. And it should be remembered, that the largest body of a ship may be so balanced in the water, that an ounce less or more of weight may leave her at the surface or sink her to the bottom. There are also certain heavy cargoes, that when the water gets at them are continually dissolving, and thereby lightening the vessel, such as salt and sugar. And as to water casks mentioned above, since the quantity of them must be great in ships of war where the number of men consume a great deal of water every day, if it had been made a constant rule to bung them up as fast as they were emptied, and to dispose the empty casks in proper situations, I am persuaded that many ships which have been sunk in engagements, or have gone down afterwards, might with the unhappy people have been saved; as well as many of those which in the last war foundered, and were never heard of. While on this topic of sinking, one cannot help recollecting the well known practice of the Chinese, to divide the hold of a great ship into a number of separate chambers by partitions tight caulked, (of which you gave a model in your boat upon the Seine) so that if a leak should spring in one of them the others are not affected by it; and though that chamber should fill to a level with the sea, it would not be sufficient to sink the vessel. We have not imitated this practice. Some little disadvantage it might occasion in the stowage is perhaps one reason, though that I think might be more than compensated by an abatement in the insurance that would be reasonable, and by a higher price taken of passengers, who would rather prefer going in such a vessel. But our seafaring people are brave, despite danger, and reject such precautions of safety, being cowards only in one sense, that of *fearing* to be *thought afraid*.

I promised to finish my letter with the last observation, but the garrulity of the old man has got hold of me, and as I may never have another occasion of writing on this subject, I think I may as well now, once for all, empty my nautical budget, and give you all the thoughts that have in my various long voyages occurred to me relating to navigation. I am sure that in you they will meet with a candid judge, who will excuse my mistakes on account of my good intention.

There are six accidents that may occasion the loss of ships at sea. We have considered one of them, that of foundering by a leak. The other five are, 1. Oversetting by sudden flaws of wind, or by carrying sail beyond the bearing. 2. Fire by accident or carelessness. 3. A heavy stroke of lightning, making a breach in the ship, or firing the powder. 4. Meeting and shocking with other ships in the night. 5. Meeting in the night with islands of ice.

To that of oversetting, privateers in their first cruise have, as far as has fallen within my knowledge or information, been more subject than any other kind of vessels. The double desire of being able to overtake a weaker flying enemy, or to escape when pursued by a stronger, has induced the owners to overmast their cruizers, and to spread too much canvas; and the great number of men, many of them not seamen, who being upon deck when a ship heels suddenly are huddled down to leeward, and increase by their weight the effect of the wind. This therefore should be more attended to and guarded against, especially as the advantage of lofty masts is problematical. For the upper sails have greater power to lay a vessel more on her side, which is not the most advantageous position for going swiftly through the water. And hence it is that vessels which have lost their lofty masts, and been able to make little more sail afterwards than permitted the ship to sail upon an even keel, have made so much way, even under jury masts, as to surprise the mariners themselves. But there is besides, something in the modern form of our ships that seems as if calculated expressly to allow their oversetting more easily. The sides of a ship instead of spreading out as they formerly did in the upper works, are of late years turned in, so as to make the body nearly round, and more resembling a cask. I do not know what the advantages of this construction are, except that such ships are not so easily boarded. To me it seems a contrivance to have less room in a ship at nearly the same expense. For it is evident that the same timber and plank consumed in raising the sides from a to b, and from d to c, would have raised them from a to e, and from d to f, fig. 9. In this form all the spaces between e, a, b, and c, d, f, would have been gained, the deck would have been larger, the men would have had more room to act, and not have stood so thick in the way of the enemy's shot; and the vessel the more she was laid down on her side, the more bearing she would meet with, and more effectual to support her, as being farther from the center. Whereas in the present form, her ballast makes the chief part of her bearing, without which she would turn in the sea almost as easily as a barrel. More ballast by this means becomes necessary, and that sinking a vessel deeper in the water occasions more resistance to her going through it. The Bermudian sloops still keep with advantage to the old spreading form. The islanders in the great Pacific ocean, though they have no large ships, are the most expert boat-sailors in the world, navigating that sea safely with their proas, which they prevent oversetting by various means. Their sailing proas for this purpose have outriggers generally to windward, above the water, on which one or more men are placed to move occasionally further from or nearer to the vessel as the wind freshens or slackens. But some have their outriggers to lee-ward, which resting on the water support the boat so as to keep her upright when pressed down by the wind. Their boats moved by oars or rather by paddles, are, for long voyages, fixed two together by cross bars of wood that keep them at some distance from each other, and so render their oversetting next to impossible. How far this may be practicable in larger vessels, we have not yet sufficient experience. I know of but one trial made in Europe, which was about one hundred years since by, Sir William Petty. He built a double vessel, to serve as a pacquet boat between England and Ireland. Her model still exists in the museum of the Royal Society, where I have seen it. By the accounts we have of her, she answered well the purpose of her construction, making several voyages; and though wrecked at last by a storm, the misfortune did not appear owing to her particular construction, since many other vessels of the common form were wrecked at the same time. The advantage of such a vessel is: That she needs no ballast, therefore swims either lighter or will carry more goods; and that passengers are not so much incommoded by her rolling: to which may be added, that if she is to defend herself by her cannon, they will probably have more effect, being kept more generally in a horizontal position, than those in common vessels. I think however that it would be an improvement of that model, to make the sides which are opposed to each other perfectly parallel, though the other sides are formed as in common thus, figure 10.

The building of a double ship would indeed be more expensive in proportion to her burthen; and that perhaps is sufficient to discourage the method.

The accident of fire is generally well guarded against by the prudent captain's strict orders against smoking between decks, or carrying a candle there out of a lanthorn. But there is one dangerous practice which frequent terrible accidents have not yet been sufficient to abolish; that of carrying store-spirits to sea in casks. Two large ships, the Seraphis and the Duke of Athol, one an East-Indiaman, the other a frigate, have been burnt within these two last years, and many lives miserably destroyed, by drawing spirits out of a cask near a candle. It is high time to make it a general rule, that all the ship's store of spirits should be carried in bottles.

The misfortune by a stroke of lightning I have in my former writings endeavoured to show a method of guarding against, by a chain and pointed rod, extending, when run up, from above the top of the mast to the sea. These instruments are now made and sold at a reasonable price by *Nairne and Co.* in London, and there are several instances of success attending the use of them. They are kept in a box, and may be run up and fixed in about five minutes, on the apparent approach of a thunder gust.

Of the meeting and shocking with other ships in the night, I have known two instances in voyages between London and America. In one both ships arrived though much damaged, each reporting their belief that the other must have gone to the bottom. In the other, only one got to port; the other was never afterwards heard of. These instances happened many years ago, when the commerce between Europe and America was not a tenth part of what it is at present, ships of course thinner scattered, and the chance of meeting proportionably less. It has long been the practice to keep a *look-out before* in the channel, but at sea it has been neglected. If it is not at present thought worth while to take that precaution, it will in time become of more consequence; since the number of ships at sea is continually augmenting. A drum frequently beat or a bell rung in a dark night, might help to prevent such accidents.

Islands of ice are frequently seen off the banks of Newfoundland, by ships going between North-America and Europe. In the day-time they are easily avoided, unless in a very thick fog. I remember two instances of ships running against them in the night. The first lost her bowsprit, but received little other damage. The other struck where the warmth of the sea had wasted the ice next to it, and a part hung over above. This perhaps saved her, for she was under great way; but the upper part of the cliff taking her foretopmast, broke the shock, though it carried away the mast. She disengaged herself with some difficulty, and got safe into port; but the accident shows the possibility of other ships being wrecked and sunk by striking those vast masses of ice, of which I have seen one that we judged to be seventy feet high above the water, consequently eight times as much under water; and it is another reason for keeping a good *look-out before*, though far from any coast that may threaten danger.

It is remarkable that the people we consider as savages, have improved the art of sailing- and rowing-boats in several points beyond what we can pretend to. We have no sailing boats equal to the flying proas of the south seas, no rowing or paddling boat equal to that of the Greenlanders for swiftness and safety. The birch canoes of the North-American Indians have also some advantageous properties. They are so light that two men may carry one of them over land, which is capable of carrying a dozen upon the water; and in heeling they are not so subject to take in water as our boats, the sides of which are lowest in the middle where it is most likely to enter, this being highest in that part, as in figure 11.

The Chinese are an enlightened people, the most antiently civilized of any existing, and their arts are antient, a presumption in their favour: their method of rowing their boats differs from ours, the oars being worked either two a-stern as we scull, or on the sides with the same kind of motion, being hung parallel to the keel on a rail and always acting in the water, not perpendicular to the side as ours are, nor lifted out at every stroke, which is a loss of time, and the boat in the interval loses motion. They see our manner, and we theirs, but neither are disposed to learn of or copy the other.

To the several means of moving boats mentioned above, may be added the singular one lately exhibited at Javelle, on the Seine below Paris, where a clumsy boat was moved across that river in three minutes by rowing, not in the water, but in the air, that is, by whirling round a set of windmill vanes fixed to a horizontal axis, parallel to the keel, and placed at the head of the boat. The axis was bent into an elbow at the end, by the help of which it was turned by one man at a time. I saw the operation at a distance. The four vanes appeared to be about five feet long, and perhaps two and a half wide. The weather was calm. The labour appeared to be great for one man, as the two several times relieved each other. But the action upon the air by the oblique surfaces of the vanes must have been considerable, as the motion of the boat appeared tolerably quick going and returning; and she returned to the same place from whence she first set out, notwithstanding the current. This machine is since applied to the moving of air balloons: An instrument similar may be contrived to move a boat by turning under water.

Several mechanical projectors have at different times proposed to give motion to boats, and even to ships, by means of circular rowing, or paddles placed on the circumference of wheels to be turned constantly on each side of the vessel; but this method, though frequently tried, has never been found so effectual as to encourage a continuance of the practice. I do not know that the reason has hitherto been given. Perhaps it may be this, that great part of the force employed contributes little to the motion. For instance, (fig. 12.) of the four paddles a, b, c, d, all under water, and turning to move a boat from X to Y, c has the most power, b nearly though not quite as much, their motion being nearly horizontal; but the force employed in moving a, is consumed in pressing almost downright upon the water till it comes to the place of b; and the force employed in moving d is consumed in lifting the water till d arrives at the surface; by which means much of the labour is lost. It is true, that by placing the wheels higher out of the water, this same labour will be diminished in a calm, but where a sea runs, the wheels must unavoidably be often dipt deep in the waves, and the turning of them thereby rendered very laborious to little purpose.

Among the various means of giving motion to a boat, that of . M. Bernoulli appears one of the most singular, which was to have fixed in the boat a tube in the form of an L, the upright part to have a funnel-kind of opening at top, convenient for filling the tube with water; which descending and passing through the lower horizontal part, and issuing in the middle of the stern, but under the surface of the river, should push the boat forward. There is no doubt that the force of the descending water would have a considerable effect, greater in proportion to the height from which it descended; but then it is to be considered, that every bucket-full pumped or dipped up into the boat, from its side or through its bottom, must have its *vis inertiæ* overcome so as to receive the motion of the boat, before it can come to give motion by its descent; and that will be a deduction from the moving power. To remedy this, I would propose the addition of another such L pipe, and that they should stand back to back in the boat thus, figure 13. The forward one being worked as a pump, and sucking in the water at the head of the boat, would draw it forward while pushed in the same direction by the force at the stern. And after all it should be calculated whether the labour of pumping would be less than that of rowing. A fire-engine might possibly in some cases be applied in this operation with advantage.

Perhaps this labour of raising water might be spared, and the whole force of a man applied to the moving of a boat by the use of air instead of water; suppose the boat constructed in this form, figure 14. A, a tube round or square of two feet diameter, in which a piston may move up and down. The piston to have valves in it, opening inwards to admit air when the piston rises; and shutting, when it is forced down by means of the lever B turning on the center C. The tube to have a valve D, to open when the piston is forced down, and let the air pass out at E, which striking forcibly against the water abaft must push the boat forward. If there is added an air-vessel F properly valved and placed, the force would continue to act while a fresh stroke is taken with the lever. The boatman might stand with his back to the stern, and putting his hands behind him, work the motion by taking hold of the cross bar at B, while another should steer; or if he had two such pumps, one on each side of the stern, with a lever for each hand, he might steer himself by working occasionally more or harder with either hand, as watermen now do with a pair of sculls. There is no position in which the body of a man can exert more strength than in pulling right upwards.

To obtain more swiftness, greasing the bottom of a vessel is sometimes used, and with good effect. I do not know that any writer has hitherto attempted to explain this. At first sight one would imagine, that though the friction of a hard body sliding on another hard body, and the resistance occasioned by that friction, might be diminished by putting grease between them, yet that a body sliding on a fluid, such as water, should have no need of nor receive any advantage from such greasing. But the fact is not disputed. And the reason perhaps may be this. The particles of water have a mutual attraction, called the attraction of adhesion. Water also adheres to wood, and to many other substances, but not to grease: On the contrary they have a mutual repulsion, so that it is a question whether when oil is poured on water, they ever actually touch each other; for a drop of oil upon water, instead of sticking to the spot where it falls, as it would if it fell on a looking-glass, spreads instantly to an immense distance in a film extremely thin, which it could not easily do if it touched and rubbed or adhered even in a small degree to the surface of the water. Now the adhesive force of water to itself, and to other substances, may be estimated from the weight of it necessary to separate a drop, which adheres, while growing, till it has weight enough to force the separation and break the drop off. Let us suppose the drop to be the size of a pea, then there will be as many of these adhesions as there are drops of that size touching the bottom of a vessel, and these must be broken by the moving power, every step of her motion that amounts to a drop's breadth: And there being no such adhesions to break between the water and a greased bottom, may occasion the difference.

So much respecting the motion of vessels. But we have sometimes occasion to stop their motion; and if a bottom is near enough we can cast anchor: Where there are no soundings, we have as yet no means to prevent driving in a storm, but by lying-to, which still permits driving at the rate of about two miles an hour; so that in a storm continuing fifty hours, which is not an uncommon case, the ship may drive one hundred miles out of her course; and should she in that distance meet with a lee shore, she may be lost.

To prevent this driving to leeward in deep water, a swimming anchor is wanting, which ought to have these properties.

1. It should have a surface so large as being at the end of a hauser in the water, and placed perpendicularly, should hold so much of it, as to bring the ship's head to the wind, in which situation the wind has least power to drive her.

2. It should be able by its resistance to prevent the ship's receiving way.

3. It should be capable of being situated below the heave of the sea, but not below the undertow.

4. It should not take up much room in the ship.

5. It should be easily thrown out, and put into its proper situation.

6. It should be easy to take in again, and flow away.

An ingenious old mariner whom I formerly knew, proposed as a swimming anchor for a large ship to have a stem of wood twenty-five feet long and four inches square, with four boards of 18, 16, 14, and 12, feet long, and one foot wide, the boards to have their substance thickened several inches in the middle by additional wood, and to have each a four inch square hole through its middle, to permit its being slipt on occasionally upon the stem, and at right angles with it; where all being placed and fixed at four feet distance from each other, it would have the appearance of the old mathematical instrument called a forestaff. This thrown into the sea, and held by a hauser veered out to some length, he conceived would bring a vessel up, and prevent her driving, and when taken in might be stowed away by separating the boards from the stem. Figure 15. Probably such a swimming anchor would have some good effect, but it is subject to this objection, that lying on the surface of the sea, it is liable to be hove forward by every wave, and thereby give so much leave for the ship to drive.

Two machines for this purpose have occurred to me, which though not so simple as the above, I imagine would be more effectual, and more easily manageable. I will endeavour to describe them, that they may be submitted to your judgment, whether either would be serviceable; and if they would, to which we should give the preference.

The first is to be formed, and to be used in the water on almost the same principles with those of a paper kite used in the air. Only as the paper kite rises in the air, this is to descend in the water. Its dimensions will be different for ships of different size.

To make one of suppose fifteen feet high; take a small spar of that length for the back-bone, AB, figure 16, a smaller of half that length CD, for the cross piece. Let these be united by a bolt at E, yet so as that by turning on the bolt they may be laid parallel to each other. Then make a sail of strong canvas, in the shape of figure 17. To form this, without waste of sail-cloth, sew together pieces of the proper length, and for half the breadth, as in figure 18, then cut the whole in the diagonal lines a, b, c, and turn the piece F so as to place its broad part opposite to that of the piece G, and the piece H in like manner opposite to I, which when all sewed together will appear as in figure 17. This sail is to be extended on the cross of figure 16, the top and bottom points well secured to the ends of the long spar; the two side points d, e, fastened to the ends of two cords, which coming from the angle of the loop (which must be similar to the loop of a kite) pass through two rings at the ends of the short spar, so as that on pulling upon the loop the sail will be drawn to its extent. The whole may, when aboard, be furled up, as in figure 19, having a rope from its broad end, to which is tied a bag of ballast for keeping that end downwards when in the water, and at the other end another rope with an empty keg at its end to float on the surface; this rope long enough to permit the kite's descending into the undertow, or if you please lower into still water. It should be held by a hauser. To get it home easily, a small loose rope may be veered out with it, fixed to the keg. Hauling on that rope will bring the kite home with small force, the resistance being small as it will then come endways.

R E M A R K S

Upon the Navigation from

NEWFOUNDLAND TO NEW-YORK,

In order to avoid the

GULPH STREAM

On one hand, and on the other the SHOALS *that lie to the Southward of*
Nantucket and of St. George's Banks.

AFTER you have passed the Banks of Newfoundland in about the 44th degree of
latitude, you will meet with nothing, till you draw near the Isle of Sables, which
we commonly pass in latitude 43. Southward of this isle, the current is found to extend
itself as far North as 41° 20' or 30', then it turns towards the E. S. E. or S. E. ¹/₄ E.

Having passed the Isle of Sables, shape your course for the St. George's Banks,
so as to pass them in about latitude 40°, because the current southward of those banks
reaches as far North as 39°. The shoals of those banks lie in 41° 35'.

After having passed St. George's Banks, you must, to clear Nantucket, form your
course so as to pass between the latitudes 38° 30' and 40° 45'.

The most southern part of the shoals of Nantucket lie in about 40° 45'. The
northern part of the current directly to the south of Nantucket is felt in about latitude
38° 30'.

By observing these directions and keeping between the stream and the shoals,
the passage from the Banks of Newfoundland to New-York, Delaware, or Virginia,
may be considerably shortened; for so you will have the advantage of the eddy current,
which moves contrary to the Gulph Stream. Whereas if to avoid the shoals you keep
too far to the southward, and get into that stream, you will be retarded by it at the
rate of 60 or 70 miles a day.

The Nantucket whale-men being extremely well acquainted with the Gulph
Stream, its course, strength and extent, by their constant practice of whaling on the
edges of it, from their island quite down to the Bahamas, this draft of that stream was
obtained from one of them, Capt. Folger, and caused to be engraved on the old chart
in London, for the benefit of navigators, by

B. FRANKLIN.

Note, The Nantucket captains who are acquainted with this stream, make their
voyages from England to Boston in as short a time generally as others take in
going from Boston to England, viz. from 20 to 30 days.

A stranger may know when he is in the Gulph Stream, by the warmth of the water,
which is much greater than that of the water on each side of it. If then he is
bound to the westward, he should cross the stream to get out of it as soon
as possible.

B. F.

(continued to map)

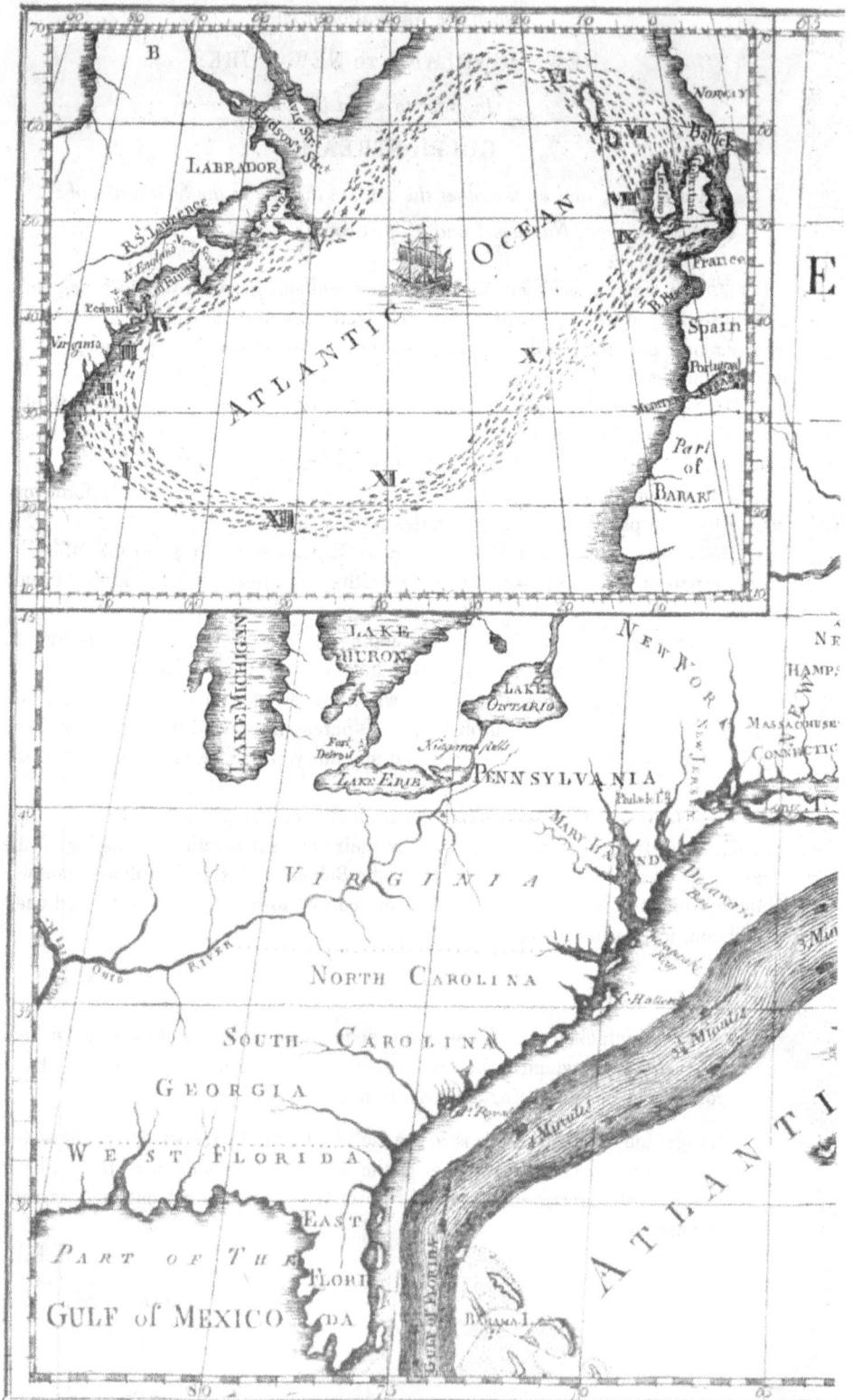

Plate

LAND
of the
SKIMAUX'S or LABRADOR
Belisle
GULF
of
NEW
St. LAURENCE
FOUNDLAND
NOVA SCOTIA
St. Jean
C. Breton
Gt. BANK
of Newfoundland
Sable I.
Baye
Cap Cod
St. Georges Bank
C. Nantucket I.
2 Minutes
2 Minutes
OCEAN
Bermuda I.
A
CHART
of The
GULF STREAM

James Poupard Sculp.

It seems probable that such a kite at the end of a long hauser would keep a ship with her head to the wind, and refining every tug, would prevent her driving so fast as when her side is exposed to it, and nothing to hold her back. If only half the driving is prevented, so as that she moves but fifty miles instead of the hundred during a storm, it may be some advantage, both in holding so much distance as is saved, and in keeping from a lee shore. If single canvas should not be found strong enough to bear the tug without splitting, it may be doubled, or strengthened by a netting behind it, represented by figure 20.

The other machine for the same purpose, is to be made more in the form of an umbrella, as represented, figure 21. The stem of the umbrella a square spar of proper length, with four moveable arms, of which two are represented C, C, figure 22. These arms to be fixed in four joint cleats, as D, D, &c. one on each side of the spar, but so as that the four arms may open by turning on a pin in the joint. When open they form a cross, on which a four square canvas sail is to be extended, its corners fastened to the ends of the four arms. Those ends are also to be flayed by ropes fastened to the stern or spar, so as to keep them short of being at right angles with it: And to the end of one of the arms should he hung the small bag of ballast, and to the end of the opposite arm the empty keg. This on being thrown into the sea, would immediately open; and when it had performed its function, and the storm over, a small rope from its other end being pulled on, would turn it, close it, and draw it easily home to the ship. This machine seems more simple in its operation, and more easily manageable than the first, and perhaps may be as effectual.*

Vessels are sometimes retarded, and sometimes forwarded in their voyages, by currents at sea, which are often not perceived. About the year 1769 or 70, there was an application made by the board of customs at Boston, to the lords of the treasury in London, complaining that the packets between Falmouth and New-York, were generally a fortnight longer in their passages, than merchant ships from London to Rhode-Island, and proposing that for the

*Captain Truxton, on board whose ship this was written, has executed this proposed machine; he has given six arms to the umbrella, they are joined to the stem by iron hinges, and the canvas is double. He has taken it with him to China. February 1786.

future they should be ordered to Rhode-Island instead of New-York. Being then concerned in the management of the American post-office, I happened to be consulted on the occasion; and it appearing strange to me that there should be such a difference between two places, scarce a day's run asunder, especially when the merchant ships are generally deeper laden, and more weakly manned than the packets, and had from London the whole length of the river and channel to run before they left the land of England, while the packets had only to go from Falmouth, I could not but think the fact misunderstood or misrepresented. There happened then to be in London, a Nantucket sea-captain of my acquaintance, to whom I communicated the affair. He told me he believed the fact might be true; but the difference was owing to this, that the Rhode-Island captains were acquainted with the gulf stream, which those of the English packets were not. We are well acquainted with that stream, says he, because in our pursuit of whales, which keep near the sides of it, but are not to be met with in it, we run down along the sides, and frequently cross it to change our side: and in crossing it have sometimes met and spoke with those packets, who were in the middle of it, and stemming it. We have informed them that they were stemming a current, that was against them to the value of three miles an hour; and advised them to cross it and get out of it; but they were too wise to be counselled by simple American fishermen. When the winds are but light, he added, they are carried back by the current more than they are forwarded by the wind: and if the wind be good, the subtraction of 70 miles a day from their course is of some importance. I then observed that it was a pity no notice was taken of this current upon the charts, and requested him to mark it out for me, which he readily complied with, adding directions for avoiding it in sailing from Europe to North-America. I procured it to be engraved by order from the general post-office, on the old chart of the Atlantic, at Mount and Page's, Tower-hill; and copies were sent down to Falmouth for the captains of the packets, who slighted it however; but it is since printed in France, of which edition I hereto annex a copy.

This stream is probably generated by the great accumulation of water on the eastern coast of America between the tropics, by the trade winds which constantly blow there. It is known that a

large piece of water ten miles broad and generally only three feet
deep, has by a strong wind had its waters driven to one side and
sustained so as to become six feet deep, while the windward side
was laid dry. This may give some idea of the quantity heaped up
on the American coast, and the reason of its running down in a
strong current through the islands into the bay of Mexico, and
from thence issuing through the gulph of Florida, and proceeding
along the coast to the banks of Newfoundland, where it turns off
towards and runs down through the Western islands. Having since
crossed this stream several times in passing between America and
Europe, I have been attentive to sundry circumstances relating to
it, by which to know when one is in it; and besides the gulph weed
with which it is interspersed, I find that it is always warmer than
the sea on each side of it, and that it does not sparkle in the night:
I annex hereto the observations made with the thermometer in two
voyages, and possibly may add a third. It will appear from them,
that the thermometer may be an useful instrument to a navigator,
since currents coming from the northward into southern seas, will
probably be found colder than the water of those seas, as the
currents from southern seas into northern are found warmer. And
it is not to be wondered that so vast a body of deep warm water,
several leagues wide, coming from between the tropics and issuing
out of the gulph into the northern seas, should retain its warmth
longer than the twenty or thirty days required to its passing the
banks of Newfoundland. The quantity is too great, and it is too
deep to be suddenly cooled by passing under a cooler air. The air
immediately over it, however, may receive so much warmth from
it as to be rarified and rise, being rendered lighter than the air
on each side of the stream; hence those airs must flow in to supply
the place of the rising warm air, and meeting with each other,
form those tornados and water-spouts frequently met with, and
seen near and over the stream; and as the vapour from a cup of
tea in a warm room, and the breath of an animal in the same room,
are hardly visible, but become sensible immediately when out in
the cold air, so the vapour from the gulph stream, in warm latitudes
is scarcely visible, but when it comes into the cool air from Newfoundland, it is condensed into the fogs, for which those parts are
so remarkable.

The power of wind to raise water above its common level in
the sea, is known to us in America, by the high tides occasioned

in all our sea-ports when a strong northeaster blows against the gulph stream.

The conclusion from these remarks is, that a vessel from Europe to North-America may shorten her passage by avoiding to stem the stream, in which the thermometer will be very useful; and a vessel from America to Europe may do the same by the same means of keeping in it. It may have often happened accidentally, that voyages have been shortened by these circumstances. It is well to have the command of them.

But may there not be another cause, independent of winds and currents, why passages are generally shorter from America to Europe than from Europe to America? This question I formerly considered in the following short paper.

On board the Pennsylvania Packet, Capt. Osborne,
At sea, April 5, 1775.

"Suppose a ship to make a voyage eastward from a place in lat. 40° north, to a place in lat. 50° north, distance in longitude 75 degrees.

"In sailing from 40 to 50, she goes from a place where a degree of longitude is about eight miles greater than in the place she is going to. A degree is equal to four minutes of time; consequently the ship in the harbour she leaves, partaking of the diurnal motion of the earth, moves two miles in a minute faster, than when in the port she is going to; which is 120 miles in an hour.

"This motion in a ship and cargo is of great force; and if she could be lifted up suddenly from the harbour in which she lay quiet, and set down instantly in the latitude of the port she was bound to, though in a calm, that force contained in her would make her run a great way at a prodigious rate. This force must be lost gradually in her voyage, by gradual impulse against the water, and probably thence shorten the voyage. Query, In returning does the contrary happen, and is her voyage thereby retarded and lengthened?"*

Would it not be a more secure method of planking ships, if instead of thick single planks laid horizontally, we were to use planks of half the thickness, and lay them double and across each

---

*Since this paper was read at the Society, an ingenious member, Mr. Patterson, has convinced the writer that the returning voyage would not, from this cause, be retarded.

other as in figure 23? To me it seems that the difference of expense would not be considerable, and that the ship would be both tighter and stronger.

The securing of the ship is not the only necessary thing; securing the health of the sailors, a brave and valuable order of men, is likewise of great importance. With this view the methods so successfully practiced by Captain Cook in his long voyages, cannot be too closely studied or carefully imitated. A full account of those methods is found in Sir John Pringle's speech, when the medal of the Royal Society was given to that illustrious navigator. I am glad to see in his last voyage that he found the means effectual which I had proposed for preserving flour, bread, &c. from moisture and damage. They were found dry and good after being at sea four years. The method is described in my printed works, page 452, fifth edition. In the same, page 469, 470, is proposed a means of allaying thirst in case of want of fresh water. This has since been practiced in two instances with success. Happy if their hunger, when the other provisions are consumed, could be relieved as commodiously; and perhaps in time this may be found not impossible. An addition might be made to their present vegetable provision, by drying various roots in slices by the means of an oven. The sweet potatoe of America and Spain, is excellent for this purpose. Other potatoes, with carrots, parsnips and turnips, might be, prepared and preserved in the same manner.

With regard to make-shifts in cases of necessity, seamen are generally very ingenious themselves. They will excuse however the mention of two or three. If they happen in any circumstance, such as after shipwreck, taking to their boat, or the like, to want a compass, a fine sewing-needle laid on clear water in a cup will generally point to the north, most of them being a little magnetical, or may be made so by being strongly rubbed or hammered, lying in a north and south direction. If their needle is too heavy to float by itself, it may be supported by little pieces of cork or wood. A man who can swim, may be aided in a long traverse by his handkerchief formed into a kite, by two cross sticks extending to the four corners; which being raised in the air, when the wind is fair and fresh, will tow him along while lying on his back. Where force is wanted to move a heavy body, and there are but few hands and no machines, a long and strong rope may make a powerful instrument. Suppose a boat is to be drawn up on a beach, that she may be out of the surf, a stake drove into the beach where you would have the boat drawn; and another to fasten the end of the rope to, which comes from the boat, and then applying what force you have to pull upon the middle of the rope at right angles with it, the power will be augmented in proportion to the length of rope between the polls. The rope being fastened to the stake A, and drawn upon in the direction C D, will slide over the stake B; and when the rope is bent to the angle A D B, represented by the pricked line in figure 24, the boat will be at B.

Some sailors may think the writer has given himself unnecessary trouble in pretending to advise them; for they have a little repugnance to the advice of landmen, whom they esteem ignorant and incapable of giving any worth notice; though it is certain that most of their instruments were the invention of landmen. At least the first vessel ever made to go on the water was certainly such. I will therefore add only a few words more, and they shall be addressed to passengers.

When you intend a long voyage, you may do well to keep your intention as much as possible a secret, or at least the time of your departure; otherwise you will be continually interrupted in your preparations by the visits of friends and acquaintance, who will not only rob you of the time you want, but put things out of your mind, so that when you come to sea, you have the mortification to recollect points of business that ought to have been done, accounts you had intended to settle, and conveniencies you had proposed to bring with you, &c. &c. all which have been omitted through the effect of these officious friendly visits. Would it not be well if this custom could be changed; if the voyager after having, without interruption, made all his preparations, should use some of the time he has left, in going himself to take leave of his friends at their own houses, and let them come to congratulate him on his happy return.

It is not always in your power to make a choice in your captain, though much of your comfort in the passage may depend on his personal character, as you must for so long a time be confined to his company, and under his direction; if he be a sensible, sociable, good natured, obliging man, you will be so much the happier. Such there are; but if he happens to be otherwise, and is only skillful, careful, watchful and alive in the conduct of his ship, excuse the rest, for these are the essentials.

Whatever right you may have by agreement in the mass of stores laid in by him for the passengers, it is good to have some particular things in your own possession, so as to be always at your own command.

1. Good water, that of the ship being often bad. You can be sure of having it good only by bottling it from a clear spring or well and in clean bottles. 2. Good tea. 3. Coffee ground. 4. Chocolate. 5. Wine of the sort you particularly like, and cyder. 6. Raisins. 7. Almonds. 8. Sugar. 9. Capillaire. 10. Lemons. 11. Jamaica spirits. 12. Eggs greas'd. 13. Diet bread. 14. Portable soup. 15. Rusks. As to fowls, it is not worth while to have any called yours, unless you could have the feeding and managing of them according to your own judgment under your own eye. As they are generally treated at present in ships, they are for the most part sick, and their flesh tough and hard as whitleather. All seamen have an opinion, broached I supposed at first prudently, for saving of water when short, that fowls do not know when they have drank enough, and will kill themselves if you give them too much, so they are served with a little only once in two days. This is poured into troughs that lie slopeing, and therefore immediately runs down to the lower end. There the fowls ride upon one another's backs to get at it, and some are not happy enough to reach and once dip their bills in it. Thus tantalized, and tormented with thirst, they cannot digest their dry food, they fret, pine, sicken and die. Some are found dead, and thrown overboard every morning, and those killed for the table are not eatable. Their troughs should be in little divisions like cups to hold the water separately, figure 25. But this is never done. The sheep and hogs are therefore your best dependance for fresh meat at sea, the mutton being generally tolerable and the pork excellent.

It is possible your captain may have provided so well in the general stores, as to render some of the particulars above recommended of little or no use to you. But there are frequently in the ship poorer passengers, who are taken at a lower price, lodge in the steerage, and have no claim to any of the cabbin provisions, or to any but those kinds that are allowed the sailors. These people are sometimes dejected, sometimes sick, there may be women and children among them. In a situation where there is no going to market, to purchase such necessaries, a few of these your superfluities distributed occasionally may be of great service, restore health, save life, make the miserable happy, and thereby afford you infinite pleasure. The worst thing in ordinary merchant ships is the cookery. They have no professed cook, and the worst hand as a seaman is appointed to that office, in which he is not only very ignorant but very dirty. The sailors have therefore a saying, that *God sends meat and the devil cooks.* Passengers more piously disposed, and willing to believe heaven orders all things for the belt, may suppose that knowing the sea-air and constant exercise by the motion of the vessel would give us extraordinary appetites, bad cooks were kindly sent to prevent our eating too much; or, that foreseeing we should have bad cooks, good appetites were furnished to prevent our starving. If you cannot trust to these circumstances, a spirit-lamp, with a blaze-pan, may enable you to cook some little things for yourself; such as a hash, a soup, & c. And it might be well also to have among your stores some potted meats, which if well put up will keep long good. A small tin oven to place with the open side before the fire, may be another good utensil, in which your own servant may roast for you a bit of pork or mutton. You will sometimes be induced to eat of the ship's salt beef, as it is often good. You will find cyder the best quencher of that thirst which salt meat or fish occasions. The ship biscuit is too hard for some sets of teeth. It may be softened by toasting. But rusk is better; for being made of good fermented bread, sliced and baked a second time, the pieces imbibe the water easily, soften immediately, digest more kindly and are therefore more wholsome than the unfermented biscuit. By the way, rusk is the true original biscuit, so prepared to keep for sea, biscuit in French signifying twice baked. If your dry peas boil hard, a two-pound iron shot put with them into the pot, will by the motion of the ship grind them as fine as mustard. The accidents I have seen at sea with large dishes of soup upon a table, from the motion of the ship, have made me wish that our potters or pewterers would make soup-dishes in divisions, like a set of small bowls united together, each containing about sufficient for one portion, in some such form as fig. 26; for then when the ship should make a sudden heel, the soup would not in a body flow over one side, and fall into people's laps and scald them, as is sometimes the case, but would be retained in the separate divisions, as in figure 27.

After these trifles, permit the addition of a few general reflections. Navigation when employed in supplying necessary provisions to a country in want, and thereby preventing famines, which were more frequent and destructive before the invention of that art, is undoubtedly a blessing to mankind. When employed merely in transporting superfluities, it is a question whether the advantage of the employment it affords is equal to the mischief of hazarding so many lives on the ocean. But when employed in pillaging merchants and transporting slaves, it is clearly the means of augmenting the mass of human misery. It is amazing to think of the ships and lives risqued in fetching tea from China, coffee from Arabia, sugar and tobacco from America, all which our ancestors did well without. Sugar employs near one thousand ships, tobacco almost as many. For the utility of tobacco there is little to be said; and for that of sugar, how much more commendable would it be if we could give up the few minutes gratification afforded once or twice a day by the taste of sugar in our tea, rather than encourage the cruelties exercised in producing it. An eminent French moralist says, that when he considers the wars we excite in Africa to obtain slaves, the numbers necessarily slain in those wars, the many prisoners who perish at sea by sickness, bad provisions, foul air, &c. &c. in the transportation, and how many afterwards die from the hardships of slavery, he cannot look on a piece of sugar without conceiving it stained with spots of human blood! Had he added the consideration of the wars we make to take and retake the sugar islands from one another, and the fleets and armies that perish in those expeditions, he might have seen his sugar not merely spotted, but thoroughly dyed scarlet in grain. It is these wars that make the maritime powers of Europe, the inhabitants of London and Paris, pay dearer for sugar than those of Vienna, a thousand miles from the sea; because their sugar costs not only the price they pay for it by the pound, but all they pay in taxes to maintain the fleets and armies that fight for it.

With great esteem, I am, Sir,

Your most obedient humble servant,

B. FRANKLIN.

Observations of the warmth of the sea-water, &c. by Fahrenheit's thermometer, in crossing the Gulph stream; with other remarks made on board the Pennsylvania packet, Capt. Osborne, bound from London to Philadelphia, in April and May 1775.

Date.	Hour.	Temp. of Air.	Temp. of Wat.	Wind.	Course.	Distance.	Latitude N.	Longitude W.	Remarks.
April 10			62						
11			61						
12			64						
13			65						
14			70						
26	8A.M.	60	70	S S E	W b S	34	37 39	60 38	Much gulph weed; saw a whale.
27	6P.M.	60	64	S W	W N W	44	37 13	62 29	Colour of water changed.
28	8A.M.	70	66	N E	W	57	37 48	64 35	No gulph weed.
—	5P.M.	67	71	NWbN	W b S	69			Sounded, no bottom.
29	11 dit.	65	72	N E	W b N	24	37 26	66 0	Much light in the water laft nig.
—	8A.M.	66	66	E S E	E b S	43			Water again of the usual deep sea colour, little or no light in it at night.
—	12	64	70	S	W b N	25	37 20	68 53	Frequent gulph weed, water continues of sea colour, little light.
30	6P.M.	62	70			60			
—	10 dit.	64	72						
—	7A.M.	65	65	S S W	W N W	44			Much light.
—	12	68	63	S W	W b N	21	38 13	72 23	Much light all laft night.
May 1	4P.M.	65	56		W N W	31			Colour of water changed.
—	10 dit.	64	56			18			
—	8A.M.	64	57	W S W	N W	18	38 43	74 3	Much light.
—	12	62	53	N W	W S W	15			Much light. Thunder-guft.
2	6P.M.	60	53	N b W	W b N	10			
—	10 dit.	64	55			30			
—	7A.M.	65	55				38 30	75 0	
3		62	54						

T t   Observations

Observations of the warmth of the sea-water, &c. by Fahrenheit's thermometer; with other remarks made on board the Reprisal, Capt. Wycks, bound from Philadelphia to France, in October and November 1776.

Date.	Hour A.M.	Hour P.M.	Temp. of Air.	Temp. of Water.	Wind.	Course.	Distance.	Latit. N.	Long. W.	Remarks.
Oct. 31	10	4	76	70 / 71	S S E	E b S	135	38 12	70 30	Left the capes Thursday night, October 29, 1776.
Nov. 1	10	4	71 / 71	78 / 81	W S W	E ¼ N	109	No ob.	68 12	
— 2	8 / 12	4	67	75 / 78	N		141	ditto.	65 23	
— 3	8 / 12	4	70 / 68	76 / 76 / 76	N W	E S E ¼ E / E b S	160	37 0	62 7	Some sparks in the water these two last nights.
— 4	9	4 / 1	68	76 / 76 / 76		N b E	194	36 26	58 8	Ditto.
— 5	8 / 12	4 / 8	68 / 70	78 / 76 / 75		N E	163	35 21	55 3	Ditto.
— 6	8 / 12	4 / 8		75 / 75 / 76	E b N	S 50 E	75	35 33	53 52	
— 7	8 / 12			77 / 78	S E b E	N 30 W	108	36 6	52 46	
— 8	9 / 12	4	75	77 / 77	S b E	N 49 E	175	38 2	50 1	
— 9	9 / 12	4	75 / 75	77 / 77 / 70	S W	N 33 E	175	39 39	46 55	

Observations

Obſervations made on board the Repriſal, continued.

Date.	Hour A.M.	Hour P.M.	Temp. of Air.	Temp. of Water.	Wind.	Courſer.	Diſtance.	Latit. N.	Long. W.	Remarks.
Novem. 9		4	70	71	E	N 17 E	64	40 39	46 27	
10	8			68						
11	12	4	56	64	S E	N 8 E	41	41 19	46 19	
12	8			63						
13	12	Noon	70	61	N N W	N 80 E	120	41 39	43 42	
14	8	4		59	E	S 84 E	69	41 29	42 10	
	all day	Noon		69	E S E	N 74 E	111	42 0	39 57	
15		4	61	68	W S W	N 70 E	186	43 3	35 51	
16	8	Noon		70	S W	N 67 W	48	43 22	34 50	
17		4	65	72	E S E	N 19 E	56	44 15	34 25	
18	8	Noon		71	S b W	N 75 E	210	45 6	29 43	
19	all day	4		69	S W	N 80 E	238	45 46	24 2	Some gulph weed.
20		Noon	65	68	N	S 80 E	155	45 19	20 30	
21	8	4		67	S	N 88 E	94	45 22	18 17	
22	9	Noon	60	67	W S W	S 89 E	133	45 9	15 19	
23	10	do.		63	N N E	S 86 E	194	45 6	10 35	
24		do.		65	N E	N 78 E	191	45 46	6 10	
25		do.		64	E	S 76 E	125	45 4	3 23	
26		do.	56	62		N 73 E	31	45 13	2 20	
27		do.		62						Soundings off Belliſle.
28			54	61						

Tt 2

1785. A Journal of a voyage from the Channel between France and England towards America.

N. B. Longitude is reckoned from London, and the Thermometer is according to Fahrenheit.

Dates.	Latit. N.		Long. W.		Therm. A.M.		Therm. P.M.		Winds.	Course.	Distance. Miles.	Variation of the Needle. West.	Therm. Noon.	
	deg	min	deg	min	Air.	Water.	Air.	Water.					A.	W.
July 29	49	15			62	57	63	58	East	S W ½ W	60	22° 0	77	78
30	48	28			62	58	62	62	E S E	W b S ¾ S	174		81	79
31	47	0			60	58	60	64	N E	S W b W	160		79	79
August 1	45	0	4	15	63	62	64	63	N W b W	S W ½ W	190		81	80
2	43	5	8	58	64	64	60	omitted	N E	S W ½ S	131		80	78
3	41	3	12	13	60	67	do.	66	N E	S W ⅓ S	166		80	79
4	38	45	15	43	66	66	65	68	N E	S S W ¼ W	165		79	77
5	36	42	17	25	67	65	71	69	N E	S S W ¼ W	149	20 0	77	75
6	35	40	19	44	70	68	68	70	N W	W S W ¼ S	137	16 30	77	75
7	35	0	21	34	72	70	73	72	North	W S W ½ S	76	11 30	80	77
8	35	51	23	10	73	71	73	74	North	S W ¾ W	112	11 15	omitted	
9	33	30	25	40	71	73	77	75	N E	W ⅓ S	143		75	74
10	33	17	27	0	74	73	76	77	S S E	W ¼ S	103		80	76
11	33	22	28	42	76	74	78	77	N E	W ½ S	50		80	76
12	33	45	31	30	76	75	78	79	W N W	S W ¼ W	35		81	78
13	34	14	33	32	78	76	81	79	West	N W ¼ N	38		78	78
14	35	37	34	31	79	79	80	80	W S W	N N W	75		78	78
15	36	7	35	0	80	79	80	78	N W b W	W N W ½ N	65		78	78
16	36	38	35	30	80	78	omitted	omitted	W S W	N W ¾ W	49		80	80
17	37	38	36	4	78	77	78	77	West	N ¼ W	62		83	80
18	37	15	37	16	78	76	omit ted	omitted	W b S	S b W	82		83	81
19	36	40	38	0	73	74	78	77	North	S S W	38		84	81
20	35	35	38	6	77	76	80	77	W N W	W ½ S	100		83	81
21	35	12	38	26	79	77	78	75	W b S	S W ¼ W	41		82	81
22	35	40	38	44	75	73	75	76	North	W N W ½ N	60	8 0	78	80
23	35	30	40	52	79	76	79	76	W N W	S W ¼ S	14		78	79
24	35	14	41	31	79	76	80	78	W b N	W S W ¾ S	38		75	79
25	34	23	42	33	79	77	75	76	S W b W	S W b S	60		78	80
26	34	12	42	44	78	76	79	79	West	W ¼ S	94		78	72
27	34	5	43	23	77	77	80	78	N N E	W ⅖ S	134			
28	34	20	44	0	78	78	81	78	N E	W ½ N	129			
29	34	20	45	52	80	78	78	78	East	W b N ¼ N	86			
30	34	55	48	31	81	78	78	79	East	W b N ⅓ N	125			
31	35	30	51	4	83	80	81	80	S S W	W ¼ N	114	6 0		
Septem. 1	35	50	52	47	83	80	omit ted	80	S W b S	W b N	82			
2	35	55	55	12	82	80	83	81	S W ¾ W	S S W	96			
3	36	20	57	24	81	81	83	80	S S W	W ½ S	75			
4	34	45	59	1	80	80	82	81	N W b N	N W N W	86			
5	35	20	61	0	87	81	78	79	N W b W	N W	74			
6	36	20	62	30	75	80	78	73	North		108			
7	34	45	63	10	77	79	75	70	N E		126			
8	34	5	64	40		79	77		E N E					
9	34	45	66	42		79								
10	37	20	68	40		73								

OBSERVATIONS

July 31. At one P. M. the Start bore W N W. distant six leagues.

August I. The water appears luminous in the ship's wake.

___2. The temperature of the water is taken at eight in the morning and at eight in the evening.

___6. The water appears less luminous.

___7. Formegas S W. dist. 32¹/₂ deg. St. Mary's SW ¹/₂ S 33 leagues.

___8. From this date the temperature of the water is taken at eight in the morning and at six in the evening.

___10. Moonlight, which prevents the luminous appearance of the water.

___11. A strong southerly current.

___12. Ditto. From this date the temperature of the air and water was taken at noon, as well as morning and evening.

___16. Northerly current.

___19. First saw gulph weed.

___21. Southerly current.

___22. Again saw gulph weed.

___24. The water appeared luminous in a small degree before the moon rose.

___29. No moon, yet very little light in the water.

___30. Much gulph weed to-day.

___31. Ditto.

Sept. 1. Ditto.

___2. A little more light in the water.

___4. No gulph weed to-day. More light in the water.

___5. Some gulph weed again.

___6. Little light in the water. A very hard thunder-gust in the night.

___7. Little gulph weed.

___8. More light in the water. Little gulph weed.

___9. Little gulph weed. Little light in the water last evening.

___10. Saw some beds of rock-weed; and we were surprised to observe the water six degrees colder by the thermometer than the preceding noon.

This day (10th) the thermometer still kept descending, and at five in the morning of the 11th, it was in water as low as 70, when we struck soundings. The same evening the pilot came on board, and we found our ship about five degrees of longitude a-head of the reckoning, which our captain accounted for by supposing our course to have been near the edge of the gulph stream, and thus an eddy-current always in our favour. By the distance we ran from Sept. 9, in the evening, till we struck soundings, we must have then been at the western edge of the gulph stream, and the change in the temperature of the water was probably owing to our suddenly passing from that current, into the waters of our own climate.

On the 14th of August the following experiment was made. The weather being perfectly calm, an empty bottle, corked very tight, was sent down 20 fathoms, and it was drawn up still empty. It was then sent down again 35 fathoms, when the weight of the water having forced in the cork, it was drawn up full; the water it contained was immediately tried by the thermometer, and found to be 70, which was six degrees colder than at the surface: The lead and bottle were visible, but not very distinctly so, at the depth of 12 fathoms but when only 7 fathoms deep, they were perfectly seen from the ship. This experiment was thus repeated Sept. 11, when we were in soundings of 18 fathoms. A keg was previously prepared with a valve at each end, one opening inward the other outward; this was sent to the bottom in expectation that by the valves being both open when going down, and both shut when coming up, it would keep within it the water received at bottom. The upper valve performed its office well, but the under one did not shut quite close, so that much of the water was lost in hauling it up the ship's side. As the water in the keg's passage upwards could not enter at the top, it was concluded that what water remained in it was of that near the ground, and on trying this by the thermometer, it was found to be at 58, which was 12 degrees colder than at the surface.

This last journal was obligingly kept for me by Mr. F. Williams, my fellow-passenger in the London Packet, who made all the experiments with great exactness.